The Open University

Science Foundation Course Unit 3

MASS, LENGTH AND TIME

Prepared by the Science Foundation Course Team

THE OPEN UNIVERSITY PRESS

A NOTE ABOUT AUTHORSHIP OF THIS TEXT

This text is one of a series that, together, constitutes *a component part* of the Science Foundation Course. The other components are a series of television and radio programmes, home experiments and a summer school.

The course has been produced by a team, which accepts responsibility for the course as a whole and for each of its components.

THE SCIENCE FOUNDATION COURSE TEAM

M. J. Pentz   (Chairman and General Editor)

F. Aprahamian	(Editor)	J. McCloy	(BBC)
A. Clow	(BBC)	J. E. Pearson	(Editor)
P. A. Crompton	(BBC)	S. P. R. Rose	(Biology)
G. F. Elliott	(Physics)	R. A. Ross	(Chemistry)
G. C. Fletcher	(Physics)	P. J. Smith	(Earth Sciences)
I. G. Gass	(Earth Sciences)	F. R. Stannard	(Physics)
L. J. Haynes	(Chemistry)	J. Stevenson	(BBC)
R. R. Hill	(Chemistry)	N. A. Taylor	(BBC)
R. M. Holmes	(Biology)	M. E. Varley	(Biology)
S. W. Hurry	(Biology)	A. J. Walton	(Physics)
D. A. Johnson	(Chemistry)	B. G. Whatley	(BBC)
A. B. Jolly	(BBC)	R. C. L. Wilson	(Earth Sciences)
A. R. Kaye	(Educational Technology)		

The following people acted as consultants for certain components of the course:

J. D. Beckman	R. J. Knight	J. R. Ravetz
B. S. Cox	D. J. Miller	H. Rose
G. Davies	M. W. Neil	
G. Holister	C. Newey	

The Open University Press
Walton Hall, Bletchley, Bucks

First Published 1971 · Reprinted 1971
Copyright © 1971 The Open University

Designed by The Media Development Group of the Open University

Printed in Great Britain by
EYRE AND SPOTTISWOODE LIMITED
AT GROSVENOR PRESS, PORTSMOUTH

SBN 335 02001 1

Open University courses provide a method of study for independent learners through an integrated teaching system, including textual material, radio and television programmes and short residential courses. This text is one of a series that make up the correspondence element of the Science Foundation Course.

The Open University's courses represent a new system of university level education. Much of the teaching material is still in a developmental stage. Courses and course materials are, therefore, kept continually under revision. It is intended to issue regular up-dating notes as and when the need arises, and new editions will be brought out when necessary.

Further information on Open University courses may be obtained from The Admissions Office, The Open University, P.O. Box 48, Bletchley, Buckinghamshire.

1.2

Contents

Table A

A List of Scientific Terms, Concepts and Principles used in Unit 3

Taken as pre-requisites			Introduced in this Unit			
1 Assumed from general knowledge	**2** Introduced in a previous Unit	Unit No.	**3** Developed in this Unit	Page No.	**4** Developed in a later Unit	Unit No.
analogue	extrapolation	1	aether	11	kinetic energy	4
	electromagnetic		order of magnitude	11		
vacuum	waves	2	special relativity	13	photon	29
	length	2	improper time	15		
sphere	model	1	proper time	15	muon	32
	muon	2	improper velocity	16		
	half-life	2	proper velocity	16	meson	32
			time dilation	17		
			frame of reference	23		
	In MAFS	**Sect.**	Newton's First Law	24		
		No.	strang	25		
	vector	4.D	Newton's Second Law	29		
	chord	2.C.3	The kilogramme	29		
			mass	29		
	addition of vectors	4.D.6	newton	30		
	Pythagoras'		newton balance	30		
	Theorem	4.C.2	scalar	33		
	cosine rule	4.D.7	weight	33		
			density	34		
	sine rule	4.D.7	centripetal force	35		
			centrifugal force	36		
			angular frequency	37		
			angular velocity	37		
			impulse	40		
			momentum	40		
			action	41		
			reaction	41		
			Newton's Third Law	42		
			conservation of momentum	43		
			relativistic mass	43		

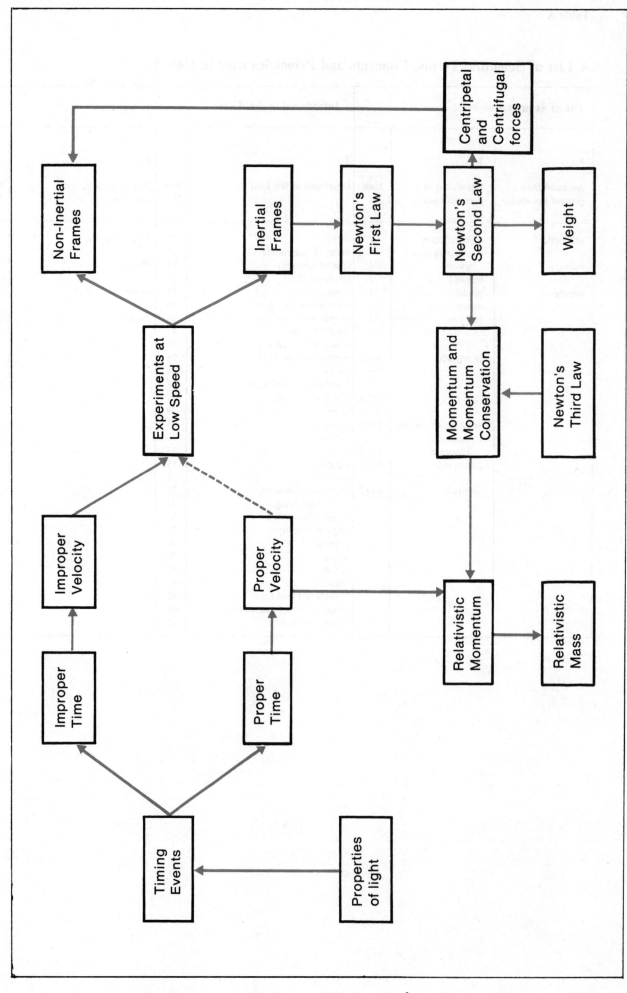

Objectives of Unit 3

When you have finished studying this Unit, you should be able to:

1 Define correctly, or recognize the best definition of, all the terms, concepts and principles in column 3 of Table A.

2 State those properties of light which are assumed in formulating the theory of special relativity. (SAQ 1)

3 Given appropriate data, to perform simple calculations relating proper time with improper time and proper velocity with improper velocity. (SAQ 2, 3, 4)

4 Perform calculations in elementary kinetics. (SAQ 6, 7, 8)

5 Perform simple calculations involving the application, with appropriate manipulation, of Newton's Laws of Motion. (SAQ 5, 9, 10 11, 12, 13, 14, 15, 16)

6 Define the momentum of a body in terms of its mass and its velocity and to be able to relate momentum with impulse. (SAQ 17, 18, 19, 20)

3.1 Introduction

In the first two Units we have pointed out some of the limitations of the human senses. We have also indicated ways in which man has attempted to extend their range. The word *attempt* is used deliberately here although, up to now, we may have given the impression that extrapolation is a pretty straightforward operation. You may, for example, have been quite satisfied with our suggested method of measuring the distance of a star (Unit 2); a method relying on an extrapolation of Earth-bound triangulation. However, we have many reservations about this technique. Only if light does travel in straight lines over large distances, only if the star itself does not move significantly during the year . . . only then are we prepared to assert that the star is so many metres away. Rather obvious qualifications, you may say. Agreed, but had we been educated within the ethos of nineteenth-century physics we might have had more serious reservations, reservations stemming from an 'intuitive' picture of how light 'ought' to behave. To begin with we shall, albeit tentatively, embrace some of these notions about the nature of light. We shall then find that our 'intuitive' picture is simply not correct. This in turn will force us to the conclusion that there is much more to seemingly obvious experiments, such as measuring the speed of a car, than we previously thought possible! It will, perhaps thankfully, emerge that there is little discrepancy between our everyday and our precise ideas of a car's velocity, *provided* that it is moving significantly slower than the velocity of light. So we will be able to discuss what we mean by velocity and how we can change the velocity of everyday objects, with, in most everyday situations, the knowledge that the qualifications are relatively unimportant. But we will have to keep the qualifications in mind. At times they will lead to very unexpected predictions.

Some of the results which we shall present may appear to run contrary to common sense. But do not forget that 'common sense' is founded on everyday experiences. In Unit 2 we showed just how limited are our senses. Therefore the rules of behaviour which applied in everyday life need not, *a priori*, be expected to hold true outside this limited range. So do not be unduly worried on meeting results which appear strange. With greater familiarity the strange becomes acceptable.

3.2 The Properties of Light

3.2.1 Collecting information

A great many experiments—some scientists would claim all—ultimately depend in one way or another on the properties of electromagnetic waves. If you attempt to measure the height of a flower with a metre stick, you must use your eyes to judge the position of the flower relative to the scale. You can use a notched rule and rely on your sense of touch if you like, but the notches must ultimately be calibrated against the wavelength of krypton light. Indeed the sense of touch itself depends on the interaction of the atoms of your fingers with the atoms of the flower. Interactions between atoms are electromagnetic. Whenever you attempt to set your watch correctly by looking at a clock or a TV screen, the correctness of the setting depends not only on the time that the light takes to travel from the screen to your eye, but on the length of time the information took in passing through the set's electronics, and on the time taken . . . right back to the moment when the hands of the studio clock were opposite particular marks on the clock face.

> If electromagnetic waves were to travel at, say, 10 m s^{-1}, by how much would the setting of your watch differ from the studio clock? In answering this question, assume that information also travels through the wires in your TV set at 10 m s^{-1}.

As an example, a set at the Open University is about 30 miles from the Oxford transmitter, which is itself about 50 miles from the studio in London, so the signal is transmitted over a total distance of about 80 miles, i.e. $80 \times 1\,500 = 1.2 \times 10^5$ m, which would take a time of $1.2 \times 10^5/10$ s $= 1.2 \times 10^4$ s, which is over $3\frac{1}{2}$ hours! (In making this type of calculation it is often helpful actually to write down that velocity=distance/time and then to cross multiply to obtain the required quantity.) Your own answer to this artificial problem should at least emphasize the importance of having a fast carrier of information when standardizing equipment. Standardization is an essential part of any quantitative investigation and of course the faster the carrier the more reliable will be the readings from the equipment that is standardized.

3.2.2 The speed of light

Why employ electromagnetic waves in standardizing equipment? Why not use sound waves instead?

> Which travel faster in air: light or sound waves?

If you do not already know the answer you might care to attempt the experiment of standing at a point equidistant from a radio set 'pipping' out the time and a TV set showing a clock face. However, as the sources of information are different, your conclusion should be suspect. The observation that in a thunderstorm the lightning flash reaches us before the thunderclap is possibly a more convincing demonstration of the answer.

If you decide that the time interval is 'zero', think again. Put a lower limit to the time interval, remembering that there is a minimum time interval of which we are conscious (a phenomenon exploited in cine photography where the apparently steady image in fact changes about 30 times each second). Your final estimate of the speed of light, or rather your estimate of the lower limit of the speed of light, will probably look pretty insignificant beside the accepted value for the velocity of light in a vacuum (or, closely enough, in air) of 2.998×10^8 m s^{-1}.

You plan to measure the speed of light by determining the time which elapses between a pulse (a short burst) of light passing two points 20 m apart. Such a pulse could be produced by a short duration spark. What sort of time intervals must the clock used in timing the pulse be able to measure?

Time intervals of $20/(3 \times 10^8)$ s, i.e. of the order of 10^{-7} s, can be measured with a cathode-ray oscilloscope. You will see such an instrument being employed in measuring the speed of light in this Unit's TV programme.

By requesting an order of magnitude measurement we are only seeking an answer which is correct to the nearest power of ten. In the present context it is of little interest whether the speed of light is, say, 576 m s^{-1} or 532 m s^{-1}, but it is worth knowing whether it is 10^{-1}, 10^0, 10^2, etc. m s^{-1}. If you have a torch you can try to measure the speed of light by flicking on the switch and measuring the time the light beam takes to travel to, say, a distant house and then return to your eyes. Alternatively switch on a light in a darkened room. You may be interested to know that Galileo attempted to measure the speed of light by placing two observers, each equipped with lamps, several miles apart. One observer uncovered his lamp at the same time starting a clock. The second observer uncovered his lamp as soon as he saw the light from the first observer's lamp. Once the first observer saw the other's lamp he stopped his clock.

3.2.3 Factors affecting the speed of light

What factors might influence the speed of light? In particular are there any which might increase the speed? If the speed of light can be increased, then not only will the rate of transmission and reception of data be accelerated but the very instruments which are employed (e.g. clocks) can be more accurately calibrated against the master instruments—the laboratory standards, as we call them.

The first experiment might be to measure the speed of light through a different medium, say water. Surface waves on a liquid are known to travel slower across, say, motor oil than paraffin. If such waves are a reasonable analogy for the way light travels, then we might expect the speed of light to depend on the medium through which it is propagated. The experiment proper can be done by measuring the time it takes for a pulse of light to travel through a glass tank containing the liquid. The results show that light travels slower in all media which transmit it than in vacuum: e.g. the speed is down by a factor of 1.33 in water, and by 1.0003 in air. So, ideally, all our information should be transmitted through a vacuum.

How does the measured speed of light depend on the speed of movement of the light source? To get a feel for possible effects you can again consider the analogous situation of ripples in water. We can even go further if we wish and tentatively adopt the nineteenth-century model that all matter, even outer space, is pervaded by a medium (traditionally called the *aether*) through which light is propagated as a wave motion. The model is only being adopted as a means of providing clues as to possible effects. If the effects are discovered it does not, of course, vindicate the model. The hypothesis that the moon is made of green cheese may be a useful one if it leads us to try the experiment of growing bacteria on moondust! If bacteria should grow it would not prove that the moon is made of green cheese!

aether

Does the speed at which ripples move across water depend on the speed of the source of the ripples? Sets of ripples can be generated by skimming a stone across water. If you prefer not to skim a stone, watch an insect flitting across the surface of a pond generating sets of ripples as it goes.

The answer, perhaps surprisingly, is that the speed of the ripples is independent of the speed of the source. But what effect does moving a light source have on the speed of the emitted light? If the speed of light were only, say, 10 m s^{-1}, it would be easy to suggest a realistic home experiment.

Suggest such an experiment.

You could repeat the experiment of bouncing a torch beam off a distant object, but this time you would move the hand holding the torch. Try the experiment if you like but we doubt whether you will be able to reach any firm conclusions. Compared to the speed of light your best efforts will look quite trivial. Perhaps the most convincing demonstration of the effect of source movement comes from experiments with sub-nuclear particles called $\pi°$ mesons. You will learn more about this type of particle in Unit 32. For the time being all you need know about the particle is that it can decay emitting in the process two short 'bursts' of radiation (called *photons*). In one such experiment $\pi°$ mesons travelling faster than 99.98 per cent of the speed of light were found, on decaying, to produce photons

At one instant

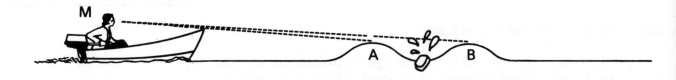

At an instant later

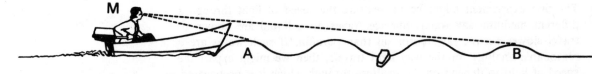

moving at a measured speed of $2.9977 \pm .0004 \times 10^8$ m s^{-1}, which is in close agreement with the speed of light (2.9979×10^8 m s^{-1}) as measured from a stationary source.

Figure 1 Showing how the apparent speed of water ripples depends on the speed of the observer. Compare the length MA in the upper drawing to the length MA in the lower one; do the same for the length MB. Satisfy yourself that MA changes much more than MB does.

Does the measured speed of water ripples depend on the speed of the observer?

It does. If you are crossing a pond towards the point where a stone has been dropped into the water, the apparent speed of the ripples coming towards you is increased; those receding from you appear to be slowed down (Fig. 1). You get the same effect if you are stationary, say on a bridge, but the water is streaming past below. What matters to an observer is his velocity relative to the medium carrying the waves.

Again, because of the high speed of light we cannot suggest realistic home experiments to look for the effect of observer movement. This remark is not, of course, intended to discourage you from running towards or away from a distant light source if you so wish! As the Earth itself is moving around the Sun at a speed of about 3×10^4 m s^{-1}, it might be more realistic to suggest experiments in which one observes a distant fixed star at six-monthly intervals; if the Earth is, say, moving towards the star in January then it will be receding from the star in July. So light from the star might appear to move faster in January than in July. However, we need not expect such effects to be readily observed for even at speed of 3×10^4 m s^{-1} the Earth is moving ten thousand times slower than light. But there are instruments called optical interferometers which can easily detect changes of 1 part in 10^6 in the velocity of light. The phenomenon of optical interference, on which these instruments depend, will be discussed in Unit 28. In 1887 two Americans, Michelson and Morley, carried out an experiment which they hoped would demonstrate the seemingly 'obvious' fact that, as with water ripples, the measured velocity of light 'must' depend on the velocity of the Earth-bound interferometer. Surprisingly, Michelson and Morley failed to find the expected dependence; the velocity of light was independent of the velocity of the observer, i.e. of the interferometer. This negative result marked the demise of the aether model. If bacteria should fail to grow on moondust the experimenter might deduce that the moon is not made of green cheese, but a more pragmatic individual would simply note that bacteria does not grow on moondust. We will simply note that the measured velocity of light is independent of the velocity of the observer. Put differently, optical interference experiments (like those carried out by Michelson and Morley) give no indication that the laboratory, or, strictly speaking, the apparatus, is moving through space. Indeed all optical experiments have failed to give evidence of any uniform motion of the laboratory through space.

velocity of light independent of velocity of observer

To sum up:

1   The velocity of light depends on the medium through which it is travelling, having the highest value in a vacuum.

2   In any laboratory the speed of light is independent of the speed of its source.

3   Optical experiments performed within a laboratory (the Earth in our case) give no information about the uniform motion of that laboratory. We shall shortly show that the same is true for mechanical experiments, as indeed it is true for all type of experiments.

Such findings as these might well have come as a distinct surprise or even a shock to someone educated during the latter half of the nineteenth century. The last two findings in particular have had, as we shall see, profound consequences; they affect the way we perform the most 'obvious' of experiments. Indeed a distinct branch of physics known as *special relativity* has built up around these two findings. Much of the credit for this is due to Einstein.

special relativity

You may now attempt *SAQ* 1, p. 54, if you wish.

3.3 Some Predictions of Special Relativity

3.3.1 Proper and improper time

The techniques involved in measuring the speed of a car from first principles may seem so obvious as scarcely to warrant description. First you measure out some distance along a road, probably by 'stepping' it out with a metre stick. Next you measure the time which elapses between the car passing the two end points. Dividing the distance by the time interval gives the car's average speed. The difficulty arises in connection with the time interval. As we shall now prove, the driver of the car measures a different time interval from that measured by an observer stationary on the road!

If we are going to measure the time taken by a car to travel a certain distance, one of the problems is reaching agreement on the exact moment when the car passes the start and finish. The simplest way to do this is probably to take separate flash photographs and then compare them at leisure. No one is going to argue about a photograph.

Figure 2 illustrates a method for obtaining our two photographs. Mounted on the car next to a clock is a flash bulb S. A mirror M at distance L above the clock is also fixed to the car. A series of clocks identical with the one on the car are set out alongside the track. Before they were set out along the road these clocks were synchronized to read the same time when together. When the bulb S is fired, the clock on the car and the adjacent clock on the road are briefly illuminated. This event is photographed by a nearby camera. To avoid possible arguments, these two clocks, the camera, and the flash bulb are kept as close as possible, i.e. they are effectively at the same point in space. Once the photograph has been taken the flash of light continues up to the mirror, is reflected, and returns to illuminate yet again the same clock on the car and, in addition, whichever new clock on the road is adjacent to the one on the car. This new event is also photographed.

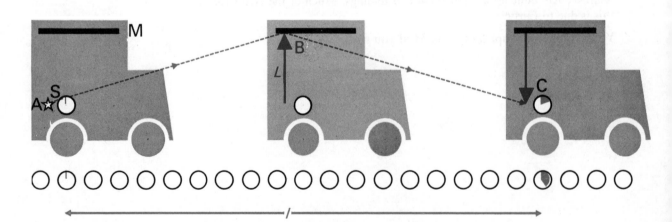

Figure 2 *A technique for measuring the speed of a car. When the flash bulb S is fired it illuminates a clock on the car and an adjacent one alongside the road. Relative to the clocks set out along the road, the light flash subsequently follows path ABC, shown dashed. Relative to the clock on the car the light proceeds vertically up to mirror M, as shown in the full line. On returning down, the light flash illuminates yet again the clock on the car and whichever new clock on the road is adjacent to the one on the car. The horizontal scale in Figure 3 is, of course, very much larger than the vertical scale. (The speed of a motor car is a lot less than the speed of light!)*

So far as the clock on the car is concerned, the light beam travels a distance 2L. Since light travels at a velocity c, the time taken for the trip, say t_{pr} (the notation will be explained shortly), is

$$t_{pr} = 2L/c \quad \ldots\ldots\ldots\ldots\ldots \quad (1)$$

proper time

and this will be the time interval as recorded by the clock on the car. A passenger on the car will therefore note that his clock records a time t_{pr} for the journey from start to finish. So far as the clocks set out along the track are concerned the light follows path ABC. The length of this path is, applying Pythagoras' theorem, $2\sqrt{L^2 + (l/2)^2}$, where l is the distance travelled by the car between the two events. Since the velocity of light is also c to the ground-based observer (remember that the Michelson-Morley experiment showed the velocity of light to be independent of the velocity of the observer), the time taken for the trip, say t_{im}, which will be the time interval recorded in the photographs of the two separated roadside clocks, is

$$t_{im} = \frac{2\sqrt{L^2 + (l/2)^2}}{c}$$

$$= \frac{2}{c}\sqrt{L^2\left[1 + \frac{1}{L^2}\left(\frac{l}{2}\right)^2\right]}$$

$$= \frac{2L}{c}\sqrt{1 + \left(\frac{l}{2L}\right)^2}$$

So from equation 1

$$t_{im} = t_{pr}\sqrt{1 + \left(\frac{l}{2L}\right)^2} \quad \ldots\ldots\ldots \quad (2)$$

improper time

The two timings are different. But how different?

Suppose the experiment is performed in a car with $L = 1.5$ m moving at 40 m s^{-1}. Remembering that light travels at 3×10^8 m s^{-1} calculate the two timings.

In this example

$$t_{pr} = \frac{2L}{c} = \frac{2 \times 1.5 \text{ m}}{3 \times 10^8 \text{ m s}^{-1}}$$

$$= 1.0 \times 10^{-8} \text{ s}$$

During this time the car will have moved a distance l given by:*

$$l = \text{velocity} \times \text{time}$$

$$= 40 \text{ m s}^{-1} \times 1.0 \times 10^{-8} \text{ s}$$

$$= 4.0 \times 10^{-7} \text{ m}$$

Substituting this value of l into equation 2 gives

$$t_{im} = t_{pr}\sqrt{1 + (l/2L)^2}$$

$$= \frac{1.0}{10^8}\sqrt{1 + \left(\frac{4.0}{10^7 \times 2 \times 1.5}\right)^2} \text{ s}$$

$$= \frac{1.0}{10^8}\sqrt{1 + \frac{1.77}{10^{14}}} \text{ s}$$

* *Strictly speaking this can only be an approximate calculation, since we are not told whether the clock used in determining the car's speed was mounted on the car or whether roadside clocks were employed. The calculation which is given is only exact if the 40 m s^{-1} was deduced by dividing the distance travelled by the car as measured along the track by the time for the journey as measured by a clock on the car.*

15

In evaluating the terms under the square root sign, we use the result obtained from the so-called binomial expansion, that when x is small compared to unity

$$(1+x)^m \approx 1 + mx + \text{second order terms which may be ignored}^* \quad \ldots\ldots\ldots\ldots\ldots\ldots (3)$$

Here $x = 1.77 \times 10^{-14}$ and $m = \frac{1}{2}$ so that,

$$t_{\text{im}} = \frac{1.0}{10^8} \left[1 + \tfrac{1}{2}\left(\frac{1.77}{10^{14}}\right) \right] \qquad \text{to a very good approximation}$$

i.e. $\quad t_{\text{im}} = 1.0000000000000089 \times 10^{-8}$ s

cf. $\quad t_{\text{pr}} = 1.0000000000000000 \times 10^{-8}$ s

In everyday life the difference between t_{im} and t_{pr} may therefore be ignored.

Try this problem (Self-Assessment Question 2, *SAQ* 2), which for revision purposes is also included with the others at the back of the Unit.

SAQ 2
Repeat the calculation on the assumption that the car is moving at a speed of 2.0×10^8 m s^{-1}. As before $L = 1.5$ m. (This speed again being deduced, using the readings of the clock on the car.)

So, when the speed of an object approaches that of light, the difference between the two timings becomes significant. Indeed, we should, in principle, always specify which time we mean, even with cars. When we use one clock, such as the clock on the car, which is present at both events, we say that we have measured a *proper time interval* (hence our notation t_{pr}). When we have two events occurring at different places, ones which could not be measured by the same clock, we say we have measured an *improper time interval* (hence t_{im}). Which of these times we care to choose in calculating the car's speed is up to us.

3.3.2 Proper and improper velocities

There are two clear ways of specifying the car's speed. One can either divide the roadside distance l by the journey time t_{pr}, as measured by a clock carried in the car, or can divide l by the journey time t_{im} measured by the roadside clocks. The first of these calculated velocities is, not surprisingly, referred to as the *proper velocity* (written v_{pr}), while the second alternative is called the *improper velocity* (written v_{im}). In symbols:

$$v_{\text{pr}} = \frac{l}{t_{\text{pr}}} \ldots\ldots\ldots\ldots\ldots\ldots (4)$$

and

$$v_{\text{im}} = \frac{l}{t_{\text{im}}} \ldots\ldots\ldots\ldots\ldots\ldots (5)$$

Can you say whether the usual car speedometer is calibrated to show proper or improper velocity?

To answer this, we must know whether, in calibrating the master speedometer (against which production speedometers are presumably checked), a clock was carried in the car, or clocks were set out alongside the road.

You will find the detailed working out of self-assessment problems at the end of the Unit, in this case on p. 57. You should follow through the worked solution if you are unsuccessful at arriving at it yourself. You should also read through the working if you obtain the correct answer but are uncertain why.

The answer you should have obtained is:

$$t_{\text{im}} = 1.20 \times 10^{-8} \text{ s}$$
$$t_{\text{pr}} = 1.00 \times 10^{-8} \text{ s}$$

The problem is worked out on page 57.

proper and improper velocity

---

* *You should satisfy yourself that when used, for example, to evaluate $(1+0.03)^2$ the binomial expansion (with x=0.03 and m=2) does lead to nearly the same answer as that obtained by squaring 1.03.*

16

At a guess it seems more likely that roadside stop-watches were employed so that a car speedometer measures improper speeds. However, as has been shown, the difference between t_{pr} and t_{im} and therefore between v_{pr} and v_{im} can be ignored in this case. But, as has also been shown, the difference in timings and therefore in velocities becomes significant when the speed of the object approaches that of light.

Having defined what is meant by velocity, we may rewrite equation 2 which relates proper and improper times, as follows:

$$t_{im} = t_{pr} \sqrt{1 + \left(\frac{l}{2L}\right)^2} \quad \ldots\ldots\ldots (2)$$

From equation 5,

$$l = v_{im}\, t_{im},$$

while from equation 1,

$$2L = ct_{pr}.$$

So

$$t_{im} = t_{pr} \sqrt{1 + \left(\frac{v_{im}t_{im}}{ct_{pr}}\right)^2}$$

Squaring both sides,

$$t^2_{im} = t^2_{pr} \left(1 + \frac{v^2_{im}t^2_{im}}{c^2 t^2_{pr}}\right)$$

$$\frac{t^2_{im}}{t^2_{pr}} = 1 + \frac{v^2_{im}t^2_{im}}{c^2 t^2_{pr}}$$

or

$$\frac{t^2_{im}}{t^2_{pr}} \left(1 - \frac{v^2_{im}}{c^2}\right) = 1$$

therefore

$$t^2_{im} = t^2_{pr} / \left(1 - \frac{v^2_{im}}{c^2}\right)$$

so

$$t_{im} = t_{pr} / \sqrt{1 - \frac{v^2_{im}}{c^2}}$$

or

$$t_{pr} = t_{im} \sqrt{1 - \frac{v^2_{im}}{c^2}} \quad \ldots\ldots\ldots\ldots (6)$$

The clock (notice the singular) on the apparatus, the one most people would consider to be 'moving', measures a smaller interval between the two events than do the other recording clocks (notice the plural). This effect is known as *time dilation*. Here is another example of dilation.

time dilation

SAQ 3
In 1965 Ron Clarke established a world record for the 10 000 metre event of 27 mins. 39.4 secs. Had Clarke carried a watch, how would his timing have differed from the judges? Needless to say the judges were stationary alongside the track.

The problem is worked out on p. 57.

Clarke's timing would have been 0.9999999999999998 of the judges, i.e. it would have been less than theirs by 2 parts in 10^{16}. Of course no present-day clock can measure times to such accuracies. So, unless there is a drastic reduction in the velocity of light (*a priori* there is no reason why the velocity of light should not change over the centuries), no physicist will query which timing is actually employed in athletic events. Should we wish to do so, we may of course express these different timings in terms of proper and improper velocity.

If in our defining relation for v_{pr} (equation 4) we substitute for t_{pr} from equation 6, we obtain

$$v_{pr} = l/t_{pr}$$
$$= l/t_{im}\sqrt{1 - v^2_{im}/c^2}$$
$$= v_{im}/\sqrt{1 - v^2_{im}/c^2} \ldots \ldots (7)$$

since v_{im} is defined as

$$v_{im} = l/t_{im} \qquad \text{(equation 5)}.$$

Equation 7 emphasizes once again that, unless the (improper) speed of an object approaches the speed of light, the difference between v_{pr} and v_{im} may safely be ignored. We shall make use of equation 7 later in the Unit.

3.3.3 Evidence for time dilation

While time dilation is never likely to be observed with large, slow-moving objects like motor cars, it does appear in a really dramatic way with small, fast-moving particles like muons. Muons break up spontaneously with a half-life of 1.53×10^{-6} s, as measured when they are at rest relative to the observer (stopped in a block of lead, for instance). This is a *proper* time.

Do you remember what half-life means?

If, for example, 1 024 muons are present at any one time, then, on average, 512 will be present after 1.53×10^{-6} (proper) seconds have passed, 256 after $2 \times 1.53 \times 10^{-6}$ s, 128 after $3 \times 1.53 \times 10^{-6}$ s, and so on.

In general, out of an initial number N, the number N_p remaining after a time t_{pr} is

$$N_p = N \times \tfrac{1}{2}^{t_{pr}/T_{\frac{1}{2}}} \ldots \ldots \ldots (8)$$

where $T_{\frac{1}{2}} = 1.53 \times 10^{-6}$ s.

This is equation 3 of Unit 2, except that we have indicated that proper times are involved.

Essentially, this experiment with fast muons consists in comparing the number of muons arriving at the top of a mountain with the number arriving at sea level.

Muons arrive with a wide range of velocities and so the apparatus at the top of the mountain is arranged so as to select only those muons with velocities within a narrow range, and the apparatus at sea level is similarly arranged to select only those muons with velocities within the same narrow range. This is done by using an arrangement of lead blocks and scintillation counters like the one you saw in the television programme of Unit 2. (See also the diagram in the broadcast notes on that programme.) With this arrangement, muons that have enough velocity to penetrate the upper lead block but not enough to penetrate the lower block as well will stop in the lower block, decay and be counted. If the upper block is very much thicker than the lower one, the range of velocities selected will be quite narrow. Since the apparatus used to select (or 'measure') the muons' velocity is at rest relative to the muons, the velocity thus measured is an *improper* velocity. (Or is it a *proper* velocity? If you have any doubt go back and read section 3.3.2 again.)

The apparatus on top of the mountain and the one at sea level are arranged to select the same range of velocities—for instance, from $0.991c$ to $0.993c$, or $(0.992 \pm 0.001)c$.

Would the upper lead block have _exactly_ the same thickness in the mountain-top apparatus as it has in the sea-level apparatus?

In a particular experiment, the velocity selected was 0.992c and the height of the mountain above sea level was 1 920 m.

Clearly, to travel down from mountain-top level to sea level, muons of this velocity took a time

$$t = 1\,920/0.992c \text{ seconds}$$
$$= 1\,920/0.992 \times 2.998 \times 10^8 \text{ s}$$
$$= 6.46 \times 10^{-6} \text{ s.}$$

Is this a _proper_ time or an _improper_ time?

It is an improper time, as it is derived from the improper velocity $v_{im} = 0.992c$. If you have any _doubts_ about this, refer back again to section 3.3.2, equation 5.

Thus, $t_{im} = 6.46 \times 10^{-6}$ s.

How many muons will have decayed in transit between the two levels?

Look again at equation 8. This tells us that if there are N muons initially then the number of muons left after a _proper_ time interval t_{pr} will be

$$N_p = N \times (\tfrac{1}{2})^{t_{pr}/T_{\frac{1}{2}}} \ldots \ldots \ldots (8)$$

where T is the _proper_ half-life time, measured, as t_{pr} is also, in a frame in which the muons are at rest.

To use equation 8 to find the answer to our problem, we have first to convert the improper time we have measured into a proper time. Refer back to equation 6.

$$t_{pr} = t_{im}\sqrt{1 - v^2{}_{im}/c^2} \ldots \ldots \ldots (6)$$
$$= 6.46 \times 10^{-6}\sqrt{1 - (0.992)^2}$$
$$= 0.815 \times 10^{-6} \text{ s.}$$

Note that the time, as measured by an observer on the Earth, is nearly eight times longer than the time as measured by the muon.

Now, the number of muons of velocity $v_{im} = (0.992 \pm 0.001)c$, counted during a certain time-interval by the mountain-top apparatus, will be a given fraction of _all_ the muons of that velocity that are passing in that time interval through the atmosphere at that altitude.

Similarly the number of muons of the same velocity, counted by the sea level apparatus during a certain time interval, will be a given fraction of all the muons of that velocity that are arriving at sea level in that time interval. We assume that the two fractions are the same, because the two 'muon sampling' arrangements are the same.

In other words, we assume that

number of muons at S $= k \times N_S$, and that

number of muons at M $= k \times N_M$

where k is the constant fraction of muons (of the given velocity) sampled by the apparatus on the mountain top, M, or by the apparatus at sea level, S (Fig. 3).

The column of air between sea level and mountain top height would itself add to the stopping power of the upper lead block in the sea-level apparatus. So one would add an _equivalent_ thickness of lead to the upper block in the mountain-top apparatus if one wanted the velocities selected at the two levels to be accurately equal.

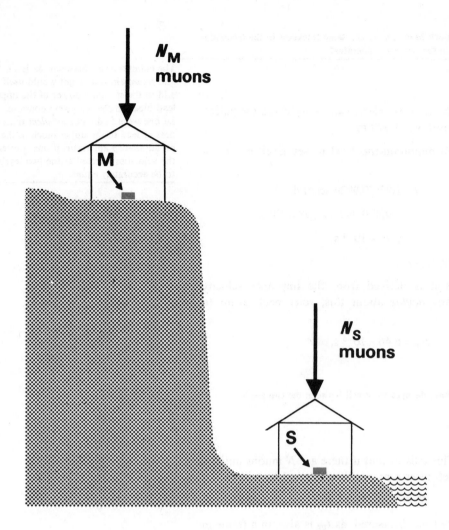

Figure 3 *An experiment demonstrating time dilation. The number of muons passing through counter M in a laboratory on top of a mountain is compared with the number passing through counter S in a laboratory at sea level. The experimentally measured ratio for these two counts agrees with the ratio to be expected, only if the muons 'keep' proper time.*

So,
$$\frac{\text{number of muons at S}}{\text{number of muons at M}} = \frac{N_S}{N_M} \quad \ldots\ldots\ldots\ldots\ldots\ldots(9)$$

In the particular experiment quoted, the ratio of counts was found to be

$$\frac{N_S}{N_M} = 0.691.$$

Does this agree with the ratio that can be predicted from equation 8? If we use the value of t_{pr} we calculated from equation 6 and the value of $T_{\frac{1}{2}}$ found from the experiment in Unit 2, then the ratio N_s/N_m in equation 8 must be the same as the ratio of muons at S to muons at M.

So

$$\frac{\text{number of muons at S}}{\text{number of muons at M}} = \frac{N_p}{N} = (\tfrac{1}{2})^{t_{pr}/T_{\frac{1}{2}}}$$

$$= (\tfrac{1}{2})^{0.815\times10^{-6}/1.53\times10^{-6}}$$

$$= (\tfrac{1}{2})^{0.533*}$$

$$= 0.691.$$

The close agreement between the predicted and measured count ratio shows that the use of equation 6 to convert the measured *improper* time to the proper time was in fact correct.

* *If you are not sure what is meant by $(\tfrac{1}{2})^{0.533}$, you should read section 1.A of MAFS. If you cannot evaluate quantities like $(\tfrac{1}{2})^{0.533}$, you should read section 1.B of MAFS.*

What result would we have predicted if we had not taken time dilation into account?

Had we used 'common sense' and decided that the flight time 'must' be the same to the muon as it is to the ground based observers, we would have used $t_{\text{lm}} = 6.46 \times 10^{-6}$ s in equation 8. This would have led us to calculate the count ratio as

$$\frac{\text{number of muons at S}}{\text{number of muons at M}} = (\tfrac{1}{2})^{6.46 \times 10^{-6}/1.53 \times 10^{-6}}$$

$$= (\tfrac{1}{2})^{4.21}$$

$$= 0.054.$$

This is some thirteen times less than the observed ratio.

Evidently, this sort of 'common sense' is not always to be trusted!

Another consequence of the difference between proper and improper time is that we must revise our ideas about the *simultaneity* of events that occur at different places. This problem is discussed, with the aid of animations, in the TV programme for this Unit.

You may now attempt SAQ 4, p. 54, if you wish.

3.4 Newton's First Law

Having discovered that there is more than meets the eye to the measurement of time intervals, we can examine in more detail the behaviour of large (or *macroscopic*) objects which are moving at speeds significantly less than the speed of light.

How fast can an object move before the discrepancy between proper and improper time reaches 1 per cent?

Substituting $t_{pr}/t_{im} = 99/100$ into equation 6 gives

$$\frac{99}{100} = \sqrt{1 - \frac{v_{im}^2}{c^2}}$$

On squaring,

$$\left(\frac{99}{100}\right)^2 = 1 - \frac{v_{im}^2}{c^2}$$

Therefore

$$\frac{v_{im}}{c} = \sqrt{1 - \left(\frac{99}{100}\right)^2}$$

$$= \sqrt{\frac{100^2 - 99^2}{100^2}}$$

or, factorizing the numerator*

$$= \sqrt{\frac{(100 + 99)(100 - 99)}{100^2}}$$

$$= \sqrt{\frac{199}{100^2}}$$

Therefore

$$v_{im} = \frac{14.1}{100} c.$$

So unless the speed of the object exceeds 14 per cent of that of light, i.e. is greater than 4.2×10^7 m s^{-1}, we can safely ignore the difference between proper and improper time, or, if you prefer, the difference between proper and improper velocity. Even with present-day space rockets moving at 10^5 m s^{-1}, it would be pedantic to enquire who held the clocks! But should we wish to speculate about how future space rockets, moving at speeds approaching that of light, might behave, or should we wish to describe the behaviour of nuclear particles, which can readily be made to move at these speeds, then we must make clear which time (or which velocity) we mean.

Here are a few simple experiments to try. Indeed you will probably consider them so simple and so predictable that you may be tempted to 'skip' them. Find something that can roll freely in any direction.

Experiment 1

Try to place a ping-pong ball, marble or whatever on a smooth level surface, so that it stays still. If it rolls away you will no doubt proceed to

To show that $x^2 - a^2 = (x + a)(x - a)$, multiply out the right-hand side. It is $x^2 - xa + ax - a^2$, which is indeed $x^2 - a^2$.

'level' the table. By your actions you are effectively eliminating any sideways pull which the Earth may exert on the object. You may now be tempted to assert that 'every object remains at rest when left alone'— a totally unwarranted generalization.

Experiment 2

If you have access to a record player, place the ball on the stationary turntable. If the turntable has a rubber mat, remove it. If the base is ribbed or the mat is fixed, use a sheet of thin card (several layers of paper will also do) to obtain a flat surface.

What happens?

Should the ball roll the table clearly needs levelling. Now lift the ball off the turntable, switch on the motor, preferably at 16 or 33 revolutions per minute and allow the ball to move in circles, keeping it (by hand) just above a fixed mark on the turntable. Finally, place the ball on the marked spot and let it move freely. In other words repeat experiment 1 but this time from the point of view of someone travelling on the turntable. Should you have no record player you can try the experiment on a flat pad which someone else rotates by hand. An icing stand for a cake is another possibility.

What happens?

Certainly the ball behaves differently in a rapidly rotating frame of reference than on *terra firma*. The words *frame of reference*, or simply *frame*, are a shorthand way of saying 'the laboratory in which the experiment is performed': in this case the laboratory is the turntable. More formally, the frame is the set of co-ordinate axes in which the behaviour of the object is represented graphically.

frame of reference

Try and deduce how the ball would appear to move to an observer standing on the turntable, i.e. how it would move in the frame of reference of the turntable. One technique you might experiment with is to dip the ball in ink before placing it on the turntable, which has been covered with a piece of graph paper. The graph showing the motion of the ball as seen in the turntable frame will be drawn automatically. Whatever technique you adopt you will probably conclude that 'in the frame of reference of the turntable a body left alone on a flat surface accelerates outwards along a radius'. Our cautionary words about over-generalizing on the basis of a single experiment (experiment 1) were clearly justified!

Experiment 3

Repeat experiments 1 and 2, this time setting the ball going at constant speed. As observed in one of these frames, but only in one of them, the ball continues to move at constant speed in a straight line.

Because of the experiments you carried out in Unit 1 you may well be suspicious of any experiment carried out with a ball. The conclusions are however quite general so long as essentially frictionless conditions prevail. The experiments may, for example, be repeated with the same results using small laboratory 'hovercraft' or 'pucks'—flat discs kept hovering by means of an air blast. In one type of apparatus, air streams through a multitude of small holes in a flat table and the discs ride on the air cushion. This is, of course, an inversion of the usual hovercraft principle. You will see such an apparatus in operation in this Unit's TV programme. Once set going in a 'fixed' laboratory (one on the Earth), a puck continues to move at constant speed in a straight line. This may best be demonstrated by

taking a time-exposure photograph while the table is illuminated by a stroboscope (a device which produces short bursts of light at fixed intervals). Figure 4 shows such a photograph. The successive images are, as you can quickly discover, equally spaced in a straight line.

We may summarize our experiments by saying that there are some frames of reference where the law that 'a body remains at rest or continues to move in a straight line at constant speed when left to itself' holds true. This law is known as *Newton's First Law of Motion*, and frames where it is true are known as *inertial frames of reference*.

Not surprisingly, frames where the law is not true are called *non-inertial*.

In which of the following 'laboratories' will Newton's First Law hold true?

(a) In a train moving at constant speed in a straight line.
(b) In a car turning a tight circle at constant speed.
(c) In a decelerating bus moving along a straight road.

An appeal to everyday experiences, such as the knowledge of what happens if you don't hold on when the bus brakes, should convince you that only (a) is inertial.

Just as optical experiments (such as for example the Michelson-Morley experiment) failed to show whether or not the laboratory was moving (at constant speed in a straight line), so have our simple mechanical tests. Newton's First Law holds true in a train 'moving' at a steady 0 m s⁻¹ (at rest), or at a steady 50 m s⁻¹. Those TV pictures of floating toothbrushes in the Apollo spacecraft are proof enough that Newton's First Law also holds in a spaceship moving at a steady 10^5 m s⁻¹.

Similarly, no biochemical test reveals uniform linear motion. A cup of tea in a station buffet tastes the same as does one made to the same recipe on an express. On the other hand a cup of tea in a violently swaying carriage, which is therefore accelerating and decelerating, seems decidedly un-inviting—our stomach knows when we are changing speed.

Our next group of experiments is going to be carried out in an inertial frame. Perhaps the most readily available inertial frame is a horizontal table on the Earth. However because the Earth rotates such a frame is strictly speaking non-inertial, but for most purposes the discrepancy is small. As you will learn in Unit 22, the discrepancy must sometimes be taken into account. In the vertical direction the Earth is not even approximately inertial; witness what happens when you release an object held above the Earth's surface.

You may now attempt *SAQ* 5, p. 54, if you wish.

Figure 4   A stroboscopically lit photograph showing a puck moving freely at constant speed along a horizontal air table. In taking this type of picture the camera shutter is left open; the successive images show where the puck was at each flash from the stroboscope. The time interval between flashes is constant.

Newton's First Law

inertial and non-inertial frames of reference

3.5 Newton's Second Law

3.5.1 Arbitrary units

It is common knowledge that to change the speed of a motor car one must step either on the brakes or on the accelerator; pushes or pulls are needed to accelerate an object. We are now going to make a systematic study of just how the motion of a body changes under the influence of a force.

The first problem is how to get a force, a reproducible force. You have probably tried stretching a spring, at some time or other, and will have got the subjective impression that it always takes the same effort to keep any one spring stretched the same amount.

Let us therefore agree, albeït tentatively, that when a particular spring is stretched a defined amount it provides a fixed reproducible force. It is up to us to choose the spring and to say how far it must be stretched to provide the unit of force. For want of a better name let us call this unit of force a *strang*; you will not find this unit listed in any table of units. It is a new unit we are temporarily adopting. Later we shall abandon the strang in favour of a unit based on a more fundamental definition of force.

strang

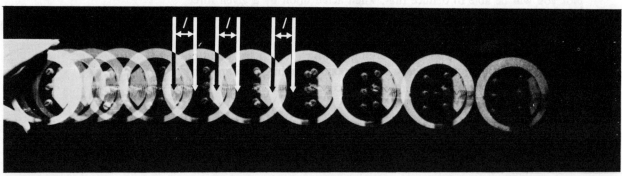

Figure 5 (a)

To investigate what a force of one strang does to a puck, the spring is attached to such a puck which is then pulled so as to stretch the spring by the agreed amount. On releasing the puck it speeds away, as expected. Figure 5a shows the result of such an experiment carried out under stroboscopic illumination. You should measure up the spring length to satisfy yourself that a constant force has been applied throughout the experiment. You should also satisfy yourself that the puck accelerates in the direction of the applied force. To study the acceleration in more detail, whether for example it is constant or not, we should plot a graph of . . .

Figure 5 The effect of applying a force of 1 strang to 1 puck. (a) A stroboscopically lit photograph of the experiment. The arrowed lines indicate the length of the spring. (b) A schematic diagram of the apparatus used to obtain Figure 5a.

What should we plot?

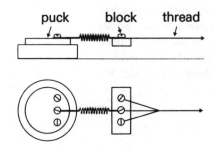

puck   block   thread

Figure 5 (b)

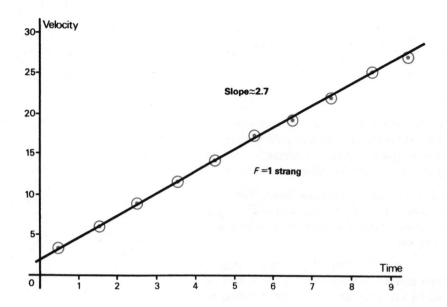

Figure 5 (c)
A graph of velocity of the puck against the time when it had this velocity. The velocity is obtained from the distance between successive images; the puck is assumed to have this velocity at a time midway between the times at which the two neighbouring images were recorded.

If you did not think of plotting a graph of velocity against time or one of distance against (time)^2 then you should now read Appendix 1. Figure 5c shows the result of making the first of these plots. From this plot you will be able to deduce that, when acted upon by a given force, the acceleration of the puck (the rate of change of its velocity) is constant, with a value given by the slope of the graph.

The following examples, detailed working of which are to be found on pp. 58–59, will allow you to check on your understanding of elementary kinetics, i.e. of the terms acceleration, velocity, distance and time, and their interrelations.

These questions assume you are familiar with the contents of Appendix 1 (Red).

SAQ 6
A car accelerates away from rest at 4 m s^{-2}. What is its velocity after it has gone 8 m?

Answer: 8 m s^{-1}. The problem is worked out on p. 58.

SAQ 7
A sprinter who is travelling at 2 m s^{-1} accelerates at a constant rate while he travels 8 m. His final velocity is 5 m s^{-1}. How long did he spend accelerating?

Answer: 2.3 s. The problem is worked out on p. 58.

SAQ 8
A puck starting from rest accelerates at a constant rate for 3 s through a distance of 5 m. What is the puck's acceleration?

Answer: 1.1 m s^{-2}. The problem is worked out on p. 59.

The next experiment we might try with the puck is to investigate the way the acceleration depends on the accelerating force. But this raises the problem of how to vary the force. Again we must resort to everyday experience—namely that it is twice as hard to keep two springs stretched when side by side as it is to keep one stretched the same amount. So we agree that if we take two parallel springs identical with the one we employed in the first investigation and stretch both the agreed amount we provide a force of two strang. With three springs the force is three strang, etc. We have now defined how forces may be added.

How do we ensure that the springs are identical?

Figure 6 (a)

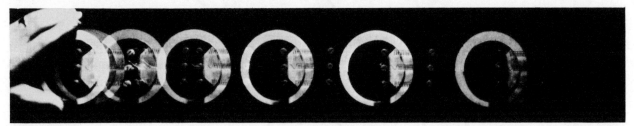

Figure 6 (b)

We should repeat the first experiment and check that all springs give the same result. Figure 6a shows the effect of applying 2 strang to one puck and Figure 6b shows the effect of 3 strang. The appropriate velocity versus time graphs are shown in Figure 6c. These plots demonstrate that with a force, F, of 2 strang the acceleration, a, is double its 1 strang value; that when F is 3 strang, a is trebled; and that, in general, the acceleration of a given body is proportional to the accelerating force:

Figure 6 The effect of applying different forces to a body of constant mass. (a) A stroboscopically lit photograph showing the effect of applying 2 strang to 1 puck. (b) The effect of 3 strang on 1 puck. (c) The corresponding graphs of velocity against time.

i.e. $\qquad a \propto F$ .... (10)

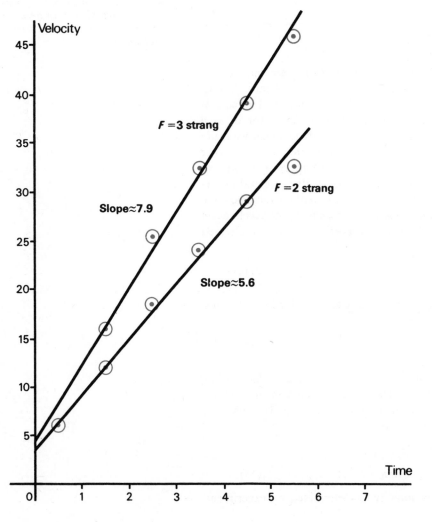

Figure 6 (c)

27

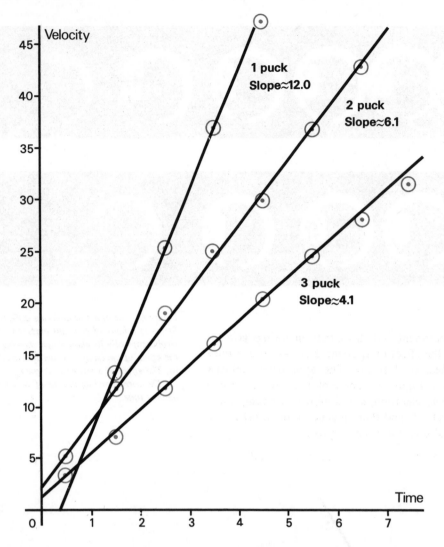

Figure 7 *Shows graphically, by means of a plot of velocity against time, the effect of applying a constant force to 1, 2 and 3 pucks.*

The next step is to vary the amount of material being accelerated (i.e. the mass, m) while keeping the accelerating force constant. Again we must agree on how to vary the mass in a controlled way. It seems obvious that the mass can be doubled by stacking two identical pucks on top of each other; trebled by stacking three identical ones, etc. (by identical we mean that each behaves in the same way when used separately in our earlier experiment). Figure 7 shows graphically the effect of applying a constant force to 1, 2 and 3 pucks. On doubling the mass, the acceleration, as measured by the slope of the graph, halves; on trebling the mass the acceleration is but a third of its one-puck value. In general, the acceleration produced by a constant force is inversely proportional to the accelerated mass.

$$a \propto \frac{1}{m} \quad \dots\dots\dots\dots\dots\dots(11)$$

The results of our two sets of experiments may be combined in the single relation

$$a \propto \frac{F}{m} \dots\dots\dots\dots\dots\dots\dots(12)$$

To satisfy yourself that equation 12 does combine the results of the separation investigations, notice that if you keep F constant in equation 12 then equation 11 is retrieved; keep m constant and equation 10 is the result. Cross-multiplying equation 12 gives

$$F \propto ma$$

or $\qquad\qquad\qquad F = Kma \dots\dots\dots\dots\dots\dots(13)$

where K, the constant of proportionality, is determined by experiment.

28

This general relation is known as *Newton's Second Law*. In one such experiment a force of 2 strang produced an acceleration of 1.38 m s^{-2} in a mass of 1 puck.

$$\therefore \quad K = \frac{F}{ma} = \frac{2}{1 \times 1.38} \quad \frac{\text{strang}}{\text{puck m s}^{-2}}$$

$$= 1.45 \text{ strang s}^2 \text{ puck}^{-1} \text{ m}^{-1}$$

The relation between force, mass and acceleration is therefore

$$F = 1.45 \, ma$$

when *F* is measured in strangs (our strangs), *m* in pucks (our pucks), and *a* in m s^{-2}.

> You decide to measure the mass of a pineapple by attaching it to a puck. A force of 3 strang gives the combined system an acceleration of 1.27 m s^{-2}. What is the mass of the pineapple?

It may not be very exciting to know that the combined mass of this pineapple and puck is $F/1.45a = 1.63$ puck, i.e. the mass of the pineapple is 0.63 puck, but this calculation should underline the fact that mass determinations are relative determinations.

> If other workers reported the relation $F = 7.8 \, ma$, would you be surprised?

Not in the least. The relation depends on what springs and pucks are used.

3.5.2 The SI units of mass and force

While there is nothing wrong with basing forces on arbitrary springs, and masses upon arbitrary pucks, it is neither a reproducible nor a permanent system. It is hard to manufacture springs with identical characteristics. In addition, these characteristics will change with age and even with the temperature of the room. A less temperamental system of units is called for. You may recall from Unit 2 how such early and variable units of length as the cubit were replaced by ever more reliable standards over the years and how today's definition of unit length is given in terms of wavelengths of a line in the spectrum of krypton. Likewise the rather variable definition of the second, in terms of the solar system, has been replaced by the definition in terms of a characteristic frequency emitted by the caesium atom. We are now going to introduce the standard of mass. From this standard of mass and those of length and time will emerge the new definition of the unit of force.

There is in a building near Paris an arbitrary lump of platinum known as The Kilogramme (kg). Originally constructed to have a mass closely equal to that of 10^3 cm^3 (10^{-3} m^3) of water at 4° C this particular lump of platinum now stands as the unit of mass in the SI system.

In imagination let us borrow The Kilogramme and place it on a frictionless surface—in practice we would of course have to use a duplicate. Next we attach an uncalibrated spring balance* (one with an unmarked scale) to The Kilogramme and pull with various steady forces until we achieve an acceleration of 1 m s^{-2}. To be exact, we should of course measure distances with some optical device using a line in the spectrum of krypton 86, and measure times (improper times in fact) with a caesium clock. Having achieved an acceleration of 1 m s^{-2}, we agree to call the arbitrary constant

The Kilogramme

---

* *The type of pocket balance used for weighing luggage might be convenient.*

force that produced this acceleration 1 kg m s^{-2}. Notice how this unit of force is defined by means of a 'thought experiment'—it only requires the kilogram, the metre and the second. There is no need for a 'standard' balance providing '1 kg m s^{-2}'. Indeed, to undertake the *practical* investigation we had to introduce another arbitrary but practical force unit—the strang. To see what value the constant of proportionality between force mass and acceleration will have in our new system of units look again at equation 13.

Deduce K in the SI system.

Not only has K the value of unity, but it is dimensionless:

$$K=\frac{F}{ma}=\frac{1\ \text{kg m s}^{-2}}{1\times1\ \text{kg m s}^{-2}}$$

Provided we are in the appropriate system, which in this case means measuring forces in kg m s^{-2}, masses in kg, and acceleration in m s^{-2},

$$F=ma \quad\dots\dots\dots\dots\dots(14)$$

As it stands this equation does not do full justice to the results of our experiment. The force, besides having a magnitude F, was applied in a definite direction. The acceleration, besides having a magnitude a, was also directed in a specific direction—the direction in which the force was applied. So the force and the acceleration should be given vectorial* representations, i.e. F and a respectively. Equation 14 should properly be written:

$$\boldsymbol{F}=m\boldsymbol{a} \quad\dots\dots\dots\dots\dots(15)$$

We may wish to calibrate an everyday spring balance in units of force. This is easily done: we simply pull a replica Kilogramme until the acceleration is say 5 m s^{-2}, when the force will be 5 kg m s^{-2}. The spot on the scale opposite the pointer can then be marked 5 kg m s^{-2}. When the acceleration is 3 m s^{-2}, the force is 3 kg m s^{-2}, etc. Since it is rather a mouthful to have to keep on saying 'kilogramme metre per second squared' as the unit of force, this is conventionally shortened to the word *newton* (written N). It is only a shortening; whenever the word 'newton' occurs as a unit of force it can be replaced by 'kg m s^{-2}'. Such a balance, calibrated to show the force it can provide in newtons, is referred to as a *newton balance*.

the newton

newton balance

There is little virtue in being able mechanically to substitute numbers into formulae to arrive at the correct solution, but a couple of examples will not be out of place if they can give you a feel for various forces you encounter in everyday life.

What push is required to give a 0.5 kg bag of sugar (i.e. one of about 1 lb.) a constant acceleration of 3 m s^{-2}?

Substituting $m=0.5$ kg and $a=3$ m s^{-2} into equation 14 gives the force required as 1.5 kg m s^{-2}, i.e. 1.5 N. It may be of some interest to ask whether an acceleration of 3 m s^{-2} would be a reasonable acceleration in practice. Suppose you did manage to heave the bag of sugar with this acceleration through a distance of 1 m (about arm's length)—would the final velocity seem plausible? You should be able to convince yourself that the final velocity is $\sqrt{2\times3\times1}\approx2.5$ m s^{-1}. (If you cannot reach this answer for yourself, re-read Appendix 1 (Red).) So the problem was a realistic one.

* *See* MAFS, *section 4.D.*

Here is another problem. This time you will have to make inspired guesses of the quantities.

SAQ 9
Make an order of magnitude estimate of the push that human legs can provide while accelerating away from rest.

The problem is worked out on p. 59.

Now a more formal problem:

SAQ 10
A car of mass 400 kg accelerates away at a constant rate from rest. In 15 s it has reached a speed of 50 km per hour. What force is the engine providing?

Answer: 370 N. The problem is worked out on p. 59.

3.5.3 Vectors

So far, in all the experiments where we deliberately set out to change the speed of an object, we have only been concerned essentially with one force. It is true that we have prescribed how parallel forces are to be added but have said nothing at all about what happens when a body is being simultaneously acted upon by several non-parallel forces. What might happen if, for example, as shown in Figure 8a, a 6 kg puck was simultaneously pulled with a force $F_1 = 5$ N in one direction, and with a force $F_2 = 8$ N at an angle of 70° to this direction (both forces in the plane of the table)?

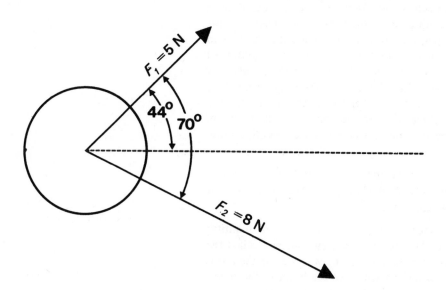

Figure 8 (a) The effect of simultaneously applying two forces to a single body. On applying a force $F_1 = 5$ N and a force $F_2 = 8$ N in the directions indicated, the puck moved along the path shown by the dashed line.

We know from our everyday experiences that an object acted upon in this manner will accelerate away in some direction between the two forces. To find out exactly what happens, all we need do is to provide two people with balances calibrated in newtons and instruct them to pull on a 6 kg puck with the specified forces in the specified directions. When this particular experiment is performed it is found that the puck accelerates away at 1.8 m s^{-2} along a linear path which lies at 44° to the 5 N force—this path is shown dashed in Figure 8(a). Now think about how someone who was unable to see the springs would describe the experiment.

Such a person would, of course, just see a 6 kg puck accelerating at 1.8 m s^{-2}. But what, quantitatively, would he infer?

Trusting equations 14 and 15, he would deduce that the puck was being acted upon by a force F of 6×1.8 kg m s^{-2}, i.e. 10.8 N, along the direction in which it accelerates. To put it differently, he could reproduce the experimental result for himself by applying a force of 10.8 N in the observed direction of motion. We, however, know that the puck is actually being simultaneously pulled by two forces. Is there any way of combining, on paper, our two pulls of 5 N and 8 N to produce a net force of 10.8 N?

Look at section 4.D on vectors and how they are added, in *MAFS*. Then try to prove that the resultant of two forces of 8 N and 5 N acting at an angle of 70° to each other is indeed a force of 10.8 N acting at an angle of 44° to the 5 N force.

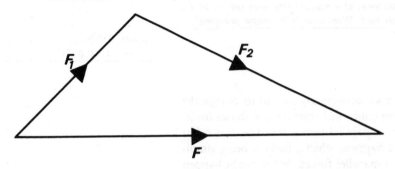

Figure 8 (b)   Shows how F$_1$ *and* F$_2$ *are added according to the 'head-to-tail' rule to produce a resultant force* F *which does indeed act along the direction in which the puck moves.*

Perhaps the easiest way to add vectors is to use the 'head-to-tail' rule, as shown in Figure 8b. First you represent one of the forces, say, as a vector, i.e. draw a line of length proportional to the magnitude of the force (on some chosen scale) and in the direction of the force. Place an arrow on this line showing the sense in which the force (F_1) acts. Next draw a similar line for F_2, placing the tail of F_2 at the head of F_1. Then join the free head and tail to get a line which represents the resultant force F. You can check for yourself that this line drawn for F does give a magnitude of 10.8 N and is in a direction of 44° to the 5 N force. You can also try drawing F_1 and F_2 in the reverse order. You will get the same result. You may prefer to calculate the resultant by applying trigonometry; the cosine rule will give the magnitude of F and the sine rule its direction. (If you are unfamiliar with the sine and cosine rules see *MAFS*, section 4 D.7.)

Although the forces appear to satisfy the 'head-to-tail' rule, this rule must be treated with caution. We shall shortly see that this rule for adding vector quantities can break down.

Velocity, e.g. the velocity of a motor car, is properly a vector quantity; it has direction and magnitude. The magnitude is customarily called the *speed*—although we shall be fairly colloquial in the way we use the terms velocity and speed, frequently using one when we really mean the other. In most situations, the head-to-tail rule makes the correct predictions when used in calculating the resultant velocity of an object, as in the following example.

SAQ **11**
An ascending escalator set at 45° to floor level is moving at 0.8 m s^{-1}. Someone is walking up the escalator at 1.1 m s^{-1}. Use the head-to-tail rule to deduce his speed and direction of movement relative to someone stationary at the foot of the escalator.

The problem is worked out on p. 60.

The predicted resultant velocity is 1.9 m s^{-1} in a direction at 45° to floor level; quite in accord with experience. But here is an example where the 'head-to-tail' rule breaks down when used in calculating the resultant velocity of an object.

> A π° meson moving at 0.9998 c disintegrates, emitting two photons. Since photons travel at a speed of c it might be argued that the resultant velocity of a photon formed when a π° meson disintegrates could be deduced by applying the head-to-tail rule, in the same way as it was applied in Q.11, namely by the velocity of the photon (of the walker) to that of the meson (of the escalator). Try applying the rule to deduce the final velocity of a photon so formed, assuming it moves in the same direction as did the meson.

The predicted value is $1.9998c$.

> But what is the observed speed of a photon emitted in such an experiment?

If you have forgotten look back at p. 12. The simple rule quite clearly breaks down when the speeds approach that of light.

3.5.4 Weight

Here is what may look like a trivial problem, but it is one that introduces a new term into your vocabulary.

> SAQ 12
> What, roughly, is the force of attraction between the Earth and a 'quarter pound' slab of chocolate? You should know that, when released, a slab falls with a constant acceleration. In an actual experiment a slab, in falling from rest through a distance of 3 m, acquired a final speed of 7.5 m s^{-1}. The problem is worked out on p. 60.

The answer is a useful one to remember; about 1 N. The pull of the Earth on an average apple is also about 1 N. Even when the slab is in our hands and is not accelerating, the downward force of attraction between it and the Earth is presumably the same. This force we call the *weight* of the object. To prevent an object accelerating downwards we must provide an upwards force equal to the weight of the object so that the resultant force is nil. The *mass* of an object is the same whether that object is on the Earth or on the Moon; the *weight* of an object is not constant. This basic difference between mass and weight is further emphasized by the different units for the two quantities: mass is measured in kilogrammes (which is a scalar* quantity), and weight in newtons (which is a vector quantity). Unfortunately for us the terms mass and weight are frequently confused in everyday speech.

weight

> An astronaut stationed on the Moon wishes to make sure that his daily sugar consumption is the same as it was on Earth. Should he measure out the same mass, or the same weight of sugar as he did on Earth?

As the astronaut clearly wishes to consume the same volume of sugar as on Earth he must take the same mass. This proportionality between the mass and the volume of a given material is actually a consequence of our definition of how masses are to be added. In section 3.5.1, for example, we took the combined mass of two identical pucks, stacked one on top of the other, to be two pucks; the general form of Newton's Second Law

---

* *Whereas a* vector *quantity has both magnitude and direction, a* scalar *quantity has magnitude only.*

(equation 14) was arrived at by making this assumption. Thus when we treble the volume of a given material, we are also trebling the mass. Put differently, we are assuming that the ratio of mass/volume of a given material is constant. This ratio, as you probably know, is called the *density* of the material.

Should the astronaut 'weigh' out the sugar, perhaps by adjusting the amount hung from the end of a spring balance until the pointer reads the same as it should on Earth, he would end up with about six times as much sugar as he required. The pull of the Moon on any one object, i.e. the weight of that object on the Moon, is only about a sixth of the weight of that object on Earth.

Properly speaking, the mass of objects can only be determined by applying a known force, F, to the object (e.g. in newton) and by measuring the resulting acceleration (e.g. in m s^{-2}). In appropriate units, as in the SI system, $m = F/a$ (equation 14). It is however possible in practice to compare two masses by 'weighing' them on a beam-balance like the one shown schematically in Figure 9. It is an experimental fact that to keep the beam of such an instrument horizontal we must apply forces F_1 and F_2 such that

$$F_1 l_1 = F_2 l_2 \ \ldots\ldots\ldots\ldots\ldots\ldots (16)$$

where l_1 and l_2 are the corresponding arm lengths.

In the balance the forces come from the 'weights', i.e. the force of gravitational attraction on the masses m_1 and m_2. Had we dropped the masses m_1 and m_2, they would have had local downward accelerations of, say, g_1 and g_2; so the forces acting on them are $m_1 g_1$ and $m_2 g_2$ respectively (from equation 15). Substituting these forces into equation 16 shows that, when the beam is horizontal,

$$m_1 g_1 l_1 = m_2 g_2 l_2 \ \ldots\ldots\ldots\ldots\ldots (17)$$

One of the really startling facts of life is that, once air resistance has been eliminated, all objects at any one spot on the Earth fall with the same acceleration, i.e. $g_1 = g_2$. This is a totally unexpected result and should thrill us more than it usually does! Knowing this result we can simplify equation 17 to

$$m_1 l_1 = m_2 l_2$$

With the traditional chemical balance, and indeed the sort of letter balance frequently seen in English village post offices, the two arms are of equal length which simplifies mass comparisons still further. It perhaps should be stressed again that such balances really measure weights; only because of the fact that all bodies fall with the same local acceleration can they be used to compare masses. Of course, in outer space, away from gravitational fields, masses can only be determined by applying a force and measuring an acceleration.

Although we shall, generally, adhere to the SI system of units, you will find us occasionally using units like pound and ton. Being units of mass these should never be used to denote a weight, although this may be excused if the word weight is attached as, e.g., in pound weight (about 4 N).

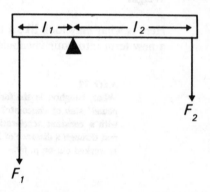

Figure 9 *A schematic diagram of a beam-balance. The beam will be horizontal when the two forces F_1 and F_2 are such that $F_1 l_1 = F_2 l_2$.*

3.5.5 Centripetal and centrifugal forces

We have now reached the stage when we can discuss the experiment of the ball placed on the turntable, the ball which spirals outwards or, according to someone travelling on the turntable, accelerates outwards in a radial direction. (Experiment 2 of section 3.4.)

Since it is easier to discuss the behaviour of an object which is travelling around in a circle than to discuss that of an object which is spiralling outwards, let us simplify the problem by constraining the object to follow a circular orbit. You need only recall childhood experience of twirling objects around on the end of a string to realize that to prevent the object from flying outwards you must exert an inward force with your hand.*

This force is called the *centripetal* force, and is denoted by F_i in Figure 10a. We are now going to derive a relation between F_i, the speed v of the object of mass m, and r, the radius of its orbit.

centripetal force

> Before reading on, try and decide how F_i depends, in a qualitative way, on m, v, and r. For example, does F_i increase as m increases? It may help to recall childhood experiments to obtain the answers although you may prefer to devise fresh experiments. You may like to make a note of what you decide and later compare it with the final deductions.
>
> Yet another way of arriving at a possible expression for F_i is by means of dimensional analysis. If you have read section 4 on 'Dimensions' in *HED*, you might like to apply this technique to deduce the relation between F_i, m, v and r. This derivation is outlined in Appendix 2 (Black).

We shall now derive an expression for F_i using what you have already learnt of Newtonian mechanics. To someone stationary on the ground the object traversing the circular path will appear at one time, say t, to be moving at a velocity v_p. This velocity v_p is shown vectorially in Figure 10a, where the tangent to the circle at P gives the instantaneous direction of movement of the mass as it passes P, while the length of the vector gives the magnitude of the velocity, i.e. the speed. At a later instant of time, which we may denote by $(t+\delta t)$,** the mass has moved to position Q, where we have indicated velocity by v_q. Although the *speed* of m is *unaltered*, the direction of movement has changed; in other words the *velocity* of m has *changed*. To find the change in velocity we merely have to discover what velocity must be added to the velocity at P to produce the value at Q.

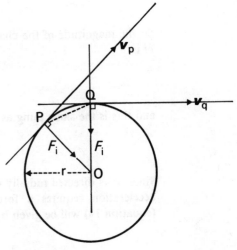

Figure 10 (a)

How would you evaluate the change in velocity of *m* as it moves from P to Q?

The rules of vector addition must be applied. This has been done in Figure 10b by means of the head-to-tail rule. Here δv denotes the velocity which must be added to the velocity v_p at P to produce the velocity v_q at Q. So far as the direction of δv is concerned, it makes equal angles with the velocities at P and Q (in triangle ABC, AB=AC). As the time interval δt gets smaller, so δv gets smaller, and the angle between the direction of δv and the tangent lines to the circle approaches 90°. Thus in the limiting case when δt approaches zero, δv is directed inwards along a radius. To find the magnitude of δv one need simply recognize that triangles OPQ and ABC are similar; similar because $\hat{POQ}=\hat{BAC}$, OP=OQ, and AB=AC.

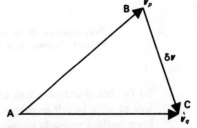

Figure 10 (b)

Figure 10 Centripetal forces. (a) In a time interval δt *a body of mass m has moved from position P to position Q. During this time interval the velocity of the mass has changed from* v_p *to* v_q. *(b) Shows how the change in velocity* δv *of the puck in moving from P to Q is obtained by applying the 'head-to-tail' rule.*

> If you cannot remember the properties of similar triangles you should read *MAFS*, section 2.E.

---

* Although the experimenter will feel an outwards force directed on him, on his hands for example, the force to consider in acceleration experiments is the force on the object which is being accelerated. When we accelerated pucks we only considered the force acting on the puck; we ignored the force which the spring exerted on the puller.

** By δt *we mean a small interval of time.*

From the properties of similar triangles

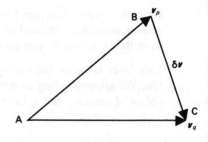

$$\frac{BC}{AB} = \frac{PQ}{OP} \quad \dots\dots\dots\dots\dots(18)$$

Now the length BC is the magnitude* of the velocity change, δv, the length AB is the magnitude* of the velocity v, and the length OP is equal to the orbit radius r. Note also that as Q approaches P, the length of the chord PQ approaches more and more closely that of the arc PQ. But the arc PQ is the path followed by the mass, moving with velocity v, in the time δt. So its length is $v\delta t$.

If we now substitute $BC = \delta v$, $AB = v$, $OP = r$ and $PQ = v\delta t$ in equation 18, we obtain

$$\frac{\delta v}{v} = \frac{v\delta t}{r}$$

So the magnitude of the change in velocity, δv, which occurs in the time δt is

$$\frac{\delta v}{\delta t} = \frac{v^2}{r}$$

and this is the same thing as the magnitude of the acceleration, a:

$$a = \frac{\delta v}{\delta t} = \frac{v^2}{r} \quad \dots\dots\dots\dots\dots(19)$$

Since δv is directed radially inwards, so is a. To produce such an inward acceleration requires a force F_i which from Newton's Second Law (equation 14) will be given by

$$F_i = \frac{mv^2}{r} \quad \dots\dots\dots\dots\dots(20)$$

—the required expression for the centripetal force. This has the same form as the relation obtained by dimensional analysis.

Does equation 20 tie in with your experimental deductions? In practice does F_i really increase as m and v increase? Does F_i increase as r decreases?

So far the discussion has centred about the ground-based observer. He will be able to tell someone on a turntable just why he, the rider, has to keep pulling inwards to constrain the object to move in a circular path. Of course the person on the turntable may not appreciate that he is in an *accelerating* frame, although his stomach (or his ears) should normally tell him so. According to the person in the rotating frame of reference the mass keeps trying to accelerate outwards as if acted upon by some invisible force, the so-called *centrifugal* force F_0. To prevent the mass from moving outwards the person on the turntable applies an inwardly directed force which he knows will equal F_0 when the radial movement ceases (no net force means no apparent acceleration).**

centrifugal force

---

* We represent the magnitude of δv by δv and the magnitude of v by v.

**The person on the turntable has, we assume, 'learnt' his mechanics while in an inertial frame. In particular, he agrees to define zero force as that which produces no apparent acceleration and furthermore that the 'head-to-tail' rule defines how forces are to be added.*

In other words he deduces that the centrifugal force F_0 is

$$F_0 = F_1 \quad \text{or, from equation 20,}$$

$$F_0 = \frac{mv^2}{r} \quad \dots\dots\dots\dots\dots(21)$$

Although it adds nothing new to the physics of the situation, there are several alternative ways of writing equations 20 and 21. If T is the time required for m to complete one circle of rotation (the *periodic time*, or, simply, the *period*), then $v = 2\pi r/T$, enabling equation 19 to be written

period of rotation

$$a = \left(\frac{2\pi r}{T}\right)^2 \Big/ r$$

$$a = r\left(\frac{2\pi}{T}\right)^2 \quad \dots\dots\dots\dots\dots(22)$$

It is customary to write $2\pi/T$ as ω (pronounced 'omega') and to call it the *angular velocity* or *angular frequency* of the particle. Making this substitution into equation 22

angular velocity
angular frequency

$$a = r\omega^2 \quad \dots\dots\dots\dots\dots(23)$$

and, as an alternative form of equation 20

$$F_1 = mr\omega^2 \quad \dots\dots\dots\dots\dots(24)$$

After deriving any relation it is always worth while pausing to see if it 'makes sense'.

SAQ 13
Calculate the force necessary to keep a 500 g bag of sugar (about 1 lb.) revolving in a circle of radius 0.75 m with a period of 0.4 s.

The detailed working of the problem is given on page 60.

Bearing in mind that the weight of an average apple is 1 N, does your answer for the force seem reasonable? Had your answer come to, say, 0.092 N, as it could have by dropping a factor of 10^3 in the calculation, would you have been worried? If not, you should try the experiment.

You should now try *SAQ*s 14–16 (p. 55).

3.6 Momentum

3.6.1 Hidden forces

It is a good idea to remind yourself at this point of the primary quality of a force, which is, at least in our experiments, to accelerate things. Indeed our first experiments were devoted to studying how the velocity of an object changed with time under the influence of a force. We could watch the velocity change. We could see the spring pulling. Now, as you will learn in later Units, there are situations where we know that the velocity of an object has changed, yet we know none of the details about how it has been changed. In such situations we may know neither the magnitude of the force nor for how long it acted. Such a situation has been created in the experiment shown in Figure 11. Here, a puck moving at constant speed enters a tunnel at a known time to emerge at a later time with a different, but again constant, speed. What, we may ask, went on in the tunnel? The change in the puck's speed indicates that a force must have acted on it while it was hidden from view. But what force, and for how long did it act?

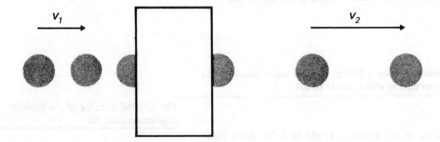

Figure 11 Showing a puck moving at constant speed entering a 'tunnel' at a known time, to emerge at a later time with a different, but again constant, speed. The diagram shows the puck at fixed time intervals.

Before we can attempt to answer these questions, we had better put down all the information we have. It is not much! We know that at one time, t_1, a puck of mass m, moving with a velocity v_1, disappeared from view only to reappear at a later time, t_2, moving at a new velocity v_2. We have

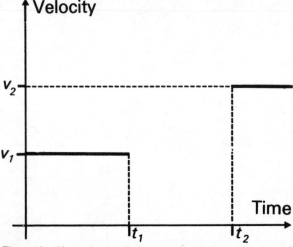

Figure 12 Showing graphically, the information contained in Figure 11, namely that at a known time, t_1, a puck moving with a speed v_1 disappeared from view to emerge at a later time t_2 moving with a speed v_2.

recorded this information in graphical form in Figure 12. Note, and this is important, the horizontal axis, the *abscissa*, represents time *not* distance; it tells us when, not where, the puck entered the tunnel.

Since we are literally in the dark as to what actually happened in the tunnel, we can only guess. Perhaps the simplest hypothesis is that during the interval t_1 to t_2 the velocity increased at a constant rate, i.e. with a constant acceleration. This guess is shown schematically in Figure 13a. If you have forgotten why a linear velocity-time graph signifies a constant acceleration, you should read Appendix 1 again. To produce a constant acceleration requires . . .

Requires what?

A constant force.

What force was acting on the puck before time t_1 and after time t_2?

None; it was moving at constant velocity. Figure 13b records the supposed forces acting on the puck. We can actually deduce the magnitude of the force, F say, which acted during the interval $t_2 - t_1$.

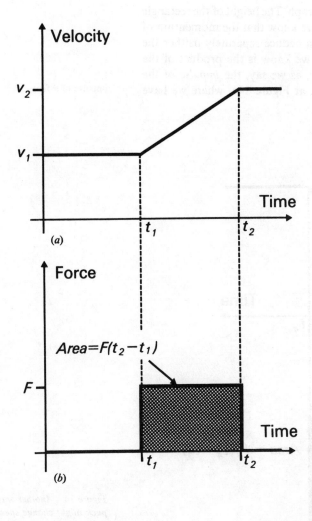

Figure 13 Shows one possible way that a puck might change speed while hidden from view. (a) Shows, graphically, the hypothesis that during the interval t_1 to t_2 the velocity increased at a constant rate. (b) The corresponding forces acting on the puck.

From equation 14

$$F = ma$$

$$= m \frac{(v_2 - v_1)}{(t_2 - t_1)}$$

Cross multiplying gives:

$$F(t_2 - t_1) = mv_2 - mv_1 \ldots\ldots\ldots\ldots (25)$$

or
$$F(t_2 - t_1) = p_2 - p_1 \ldots\ldots\ldots\ldots (26)$$

where we have written p for mv; this product mv is called the *momentum* of the object. Since velocity is a vector, the product mv is a vector quantity. If we wish to emphasize that we are dealing with vectors, equations 25 and 26 should be written as

momentum

$$\boldsymbol{F}(t_2 - t_1) = m\boldsymbol{v}_2 - m\boldsymbol{v}_1$$

or
$$\boldsymbol{F}(t_2 - t_1) = \boldsymbol{p}_2 - \boldsymbol{p}_1$$

Since the present discussion is limited to the situation where the object emerged from the tunnel moving in the same direction as when it entered the tunnel, the notation of equations 25 and 26 will be adequate.

What does the left-hand side of equation 25 or equation 26 represent in the force-time graph, Figure 13b?

It represents the area under the force-time graph. The height of the rectangle is F while the base is $(t_2 - t_1)$. So although we know that the momentum of the object changes from mv_1 to mv_2 we can deduce separately neither the force nor the time for which it acted. All we know is the product of the unknown force and the unknown time, or, as we say, the *impulse* of the force. To emphasize this uncertainty, look at Figure 14a, where we have

impulse of a force

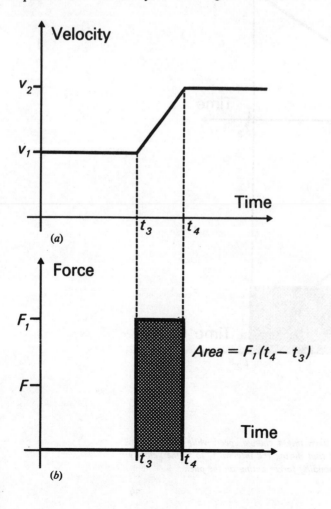

(a)

Area = $F_1(t_4 - t_3)$

(b)

Figure 14 *Another way in which the puck might change speed while hidden from view. (a) The acceleration is constant and occurs in the interval t_3 to t_4. (b) The corresponding forces acting on the puck.*

assumed that the object is again accelerated at a constant rate but the acceleration is more rapid and for a shorter time. The constant force F_1 required is shown in Figure 14b.

> **Follow through exactly the same argument as before to deduce the value of the impulse of the new constant force, F_1, acting for time $(t_4 - t_3)$.**

Once more $F_1 (t_4 - t_3) = mv_2 - mv_1$. Again all we know is the value of the impulse; we can separately infer neither F_1 nor $(t_4 - t_3)$.

> **A 0.5 kg puck enters a tunnel moving at a speed of 2.0 m s^{-1} and emerges with a speed of 8 m s^{-1}. What constant forces might have acted on the puck and for how long? Give a few examples.**

The change in momentum of the puck is

$$mv_2 - mv_1 = 0.5 \times 8.0 - 0.5 \times 2.0$$
$$= 3 \text{ kg m s}^{-1}$$

which equals the impulse of the force. So the force could have been for example 3 N for 1 s, 300 N for 10^{-2} s, or even 6×10^6 N for 5×10^{-7} s. There is no way of telling. Incidentally, you should convince yourself that the dimensions of impulse (N s) are the same as those of momentum (kg m s^{-1}). If they are not the same there is a mistake somewhere in the argument. (See section 4 in *HED*.)

Although we have assumed that the hidden force is constant, it is a general truth that, irrespective of how the force varies, the area under the force-time curve always equals the change in the object's momentum. If you are familiar with the ideas of calculus you may be able to prove this for yourself. The argument is to be found in Appendix 3 (Black).

The following example may make clear the practical importance of momentum changes and the associated impulses.

> *SAQ 17*
>
> A 90 kg person is a passenger in a car travelling at 60 km per hour, which has a head-on collision and stops 'dead'. If the car seat belt holding the passenger 'gives' for 0.2 s calculate the force which the belt exerts on the passenger as he is brought to rest.
>
> **Assume the belt provides a constant force.**

The detailed working of the problem is given on p. 61.

The answer is 7 500 N. Such a force, when applied, as it is, over a considerable area of the body is unlikely to cause significant damage. With no safety belt the dashboard may have to provide the impulse to reduce the passenger's momentum. Since a dashboard, or indeed an over-rigid seat belt, will have less 'give', the force exerted on the passenger will be larger. If the time for which the force acts is reduced say a thousand fold, its magnitude will be increased a thousand fold; the products of force and time being constant.*

---

* Although this example of the car safety belt is frequently employed as a means of giving one a 'feel' for momentum, it actually fails to distinguish momentum from kinetic energy. As you will learn in Unit 4, if a constant force F acts through a distance s, on a body of mass m, moving at a speed v, so as to bring the body to rest, then $Fs = \frac{1}{2} mv^2$. (The product $\frac{1}{2} mv^2$ is called the kinetic energy of the body.) A seat belt which 'gives' will stretch through a large distance s and so provide a small force F, such that the product Fs is equal to $\frac{1}{2} mv^2$. The damage sustained by a person who is involved in an accident depends both on his momentum and on his kinetic energy. In fact there is no way of describing what momentum is in 'everyday speech'. Most of the descriptions more closely match $\frac{1}{2} mv^2$ than mv. Before long you will meet other concepts which like momentum can only be adequately described in mathematical symbols.

3.6.2 Action-reaction forces and momentum conservation

If you have ever banged your fist on a table you will have noticed that whenever you bang the table it bangs you. You *act* on the table with a certain force and it *reacts* back with an apparently equal force on you. Place a book on a flat table and leave it alone; it leaves you alone—*action* equals *reaction*. Push on the book, i.e. act on the book, and it will push back on your fingers. Indeed it is the book's reaction which you sense. Push harder, so accelerating the book still more rapidly, and the reaction will be greater. Our subjective impression is that the reaction apparently equals the action. You might think we could prove the hunch by inserting two spring balances, one after another, between ourselves and the book. It is true that the balances give the same reading for the force but, how can a balance tell which end is being pushed, or is it pulled? We do not know whether a balance is measuring 'action' or 'reaction'. So although we cannot prove our hunch directly it does seem to be reasonable to assume that action equals reaction. This assumption is known as *Newton's Third Law*. As we shall see the prediction that momentum is reversed in collisions, a prediction made assuming that action equals reaction, is in accord with experiment. So Newton's Third Law may be verified indirectly.

action and reaction

Newton's Third Law

To see the consequences of action-reaction forces, let us consider a head-on collision between two magnetic pucks. These are pucks made of ring

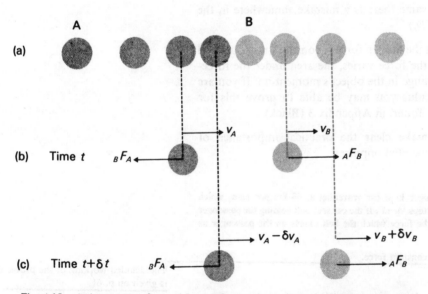

Figure 15   Action-reaction forces. (a) A puck A is shown closing in on an initially more slowly moving puck B. (b) The forces acting on each puck and the corresponding velocities at time t. *(c) The situation at a later time* t + δt.

magnets which repel each other over appreciable distances. Our argument in no way depends on the pucks being magnetic; we employ them solely to produce a gentler collision. (You will see a collision of this kind in this Unit's TV programme.) Figure 15a shows puck A, of mass m_A, closing in on an initially more slowly moving puck B, of mass m_B. If we think about the situation at a time t, as shown in Figure 15b, A will be acting on B with a force which we may write as $_AF_B$ (meaning A's push on B) while B will be reacting back on A with force $_BF_A$ (meaning, of course, B's push on A). The effect of $_AF_B$ will be to speed up B, while $_BF_A$ will slow down A. Anyone who has ever bumped into another person will know this much! To find out more about the changes consider the situation at a slightly later time, say $t + δt$, shown in Figure 15c. During the short time interval $δt$ both $_AF_B$ and $_BF_A$ will be effectively constant; their effect will have been to increase the velocity of B by $δv_B$, say, and to decrease that of A by $δv_A$.

Can you write down expressions relating the momentum changes in A and B to the impulses?

Applying equation 25 we find that

$$_BF_A\,\delta t = m_A\delta v_A \quad \dots\dots\dots\dots\dots(27)$$

$$_AF_B\,\delta t = m_B\delta v_B \quad \dots\dots\dots\dots\dots(28)$$

In view of Newton's Third Law, the left-hand sides of equations 27 and 28 are equal.

Therefore, $\qquad\qquad m_A\delta v_A = m_B\delta v_B$

As mass and the product of change in velocity is just the change in momentum we can write

$$\delta p_A = \delta p_B$$

Remembering that B has speeded up while A has slowed down, we have proved that the momentum gained by B equals that lost by A, or taken together the total momentum of A and B has not changed during the time between t and $t + \delta t$. As there is nothing special about the time interval we have examined, the same argument may be applied throughout the collision sequence. The total momentum before the collision should equal the total momentum after the collision, i.e. momentum should be *conserved*. But is it?

conservation of momentum

Find two objects of the same mass, such as a couple of similar paperback books. Place one of these on a smooth flat surface, e.g. a laminate-covered kitchen table. Give the other a shove, so that it collides head-on with the stationary one. Notice how they swop places; the moving one stops dead after the collision, the one that was stationary before moves off with the velocity which the incoming one had at the moment of impact. If you are familiar with the novelty called 'Newton's Cradle', the result will not be new to you. A billiard player also knows that when a billiard ball collides head-on with a stationary one the two balls exchange roles. Remembering that momentum is properly a vector quantity, give a vector representation to the experiment, showing that the total vector momentum before and after the collision are the same.

Your vector representation should look like that of Figure 16 although the directions may be different depending on your line of shot. If your collision is not head-on the tracks will make finite angles with each other. But even in the case of oblique collisions momentum is conserved—provided momentum is given its proper vectorial representation, i.e. vectors are drawn of magnitude mv and in the direction of v.

Figure 16 is on p. 44.

What happens to the momentum of a sliding book when it eventually slows down and stops?

Suppose the books had been projected in a West to East direction, i.e. in the same sense as the Earth's rotation.

Horizontal forces of friction between the table and the book are acting to slow the book down. At the same time, the book is reacting back on the table with horizontal forces which go to speed up the table and hence speed up the Earth's rate of rotation. Momentum is transferred from the book to the Earth.

Where did the book's momentum come from in the first place?

As you pushed on the book, it pushed back on you through your arms, to your feet, to the Earth. As you pushed the book forward it pushed the

Earth backwards. The momentum gained by the book presumably equals that lost by the Earth (as its rate of forward rotation was reduced). By the time the book had stopped moving it had given its momentum back to the Earth.

Assuming both Newton's Second and Third Laws, the conservation of momentum has been predicted. Experimentally, momentum has been found to be conserved. Since Newton's Second Law has been established independently it is tempting to conclude that Newton's Third Law is *therefore* universally correct. There could of course be other ways of accounting for momentum conservation.

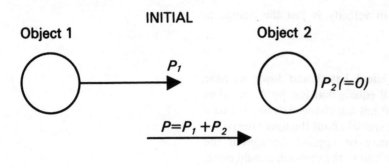

INITIAL

Object 1   Object 2

P_1 $P_2 (=0)$

$P = P_1 + P_2$

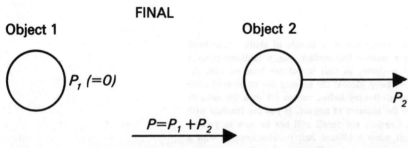

FINAL

Object 1   Object 2

$P_1 (=0)$

P_2

$P = P_1 + P_2$

Figure 16 Illustrating momentum conservation in a head-on collision between two objects of the same mass, one of which was initially at rest. p_1 represents the momentum of the incoming object and p_2 (=0) the momentum of the object at rest. Underneath is shown the total initial momentum, p. As a result of the collision the incoming object has come to rest ($p_1 = 0$) while the object initially at rest moves off with momentum p_2 equal to that of the incoming object. The total final momentum is the same as the total initial momentum.

3.6.3 Relativistic momentum

Throughout this discussion of momentum we have failed to state whether we have been talking about proper or improper velocity, or, if you prefer, whether the clocks used in measuring the velocities of the objects travel with the objects, or are stationary beside the scale measuring the distance travelled. As we saw in section 3.3, such carelessness makes little difference unless the speeds approach that of light. But once we attempt to discuss the behaviour of such fast moving objects, we must state clearly which velocity is meant. Normally, of course, improper velocities are measured; in any practical set up the clocks will be alongside the scale. If we check up on whether momentum, as defined by mass × improper velocity is conserved in collisions between, say, sub-nuclear particles moving at speeds approaching that of light, we find that the law of conservation of momentum breaks down. However, the conservation law does still hold *provided* momentum is defined as

$$p = m_0 v_{1\mathrm{m}} / \sqrt{1 - (v_{1\mathrm{m}}^2 / c^2)} \dots \dots \dots \dots (29)$$

where m_0, known as the *rest mass* of the object, is the mass that is measured at very low speeds, i.e. what, up to now, we have been calling m. At such low speeds, where $v_{1\mathrm{m}}$ is very much less than c, the ratio $v_{1\mathrm{m}}/c$ is negligible

rest mass

44

compared to unity and equation 29 reduces to $m_0 v_{\text{lm}}$, which is our original definition of momentum. (Remember, in our previous formulation of mv, m was the mass at low speeds which should be written m_0, and v was actually v_{lm}.)

There are two ways of interpreting equation 29. One way is to combine the denominator with the v_{lm} factor in the numerator, thus:

$$p = m_0 \times \frac{v_{\text{lm}}}{\sqrt{1 - v^2{}_{\text{lm}}/c^2}}$$

Now, from equation 6

$$v_{\text{pr}} = \frac{v_{\text{lm}}}{\sqrt{1 - v^2{}_{\text{lm}}/c^2}} \cdots \cdots (6)$$

In this way, momentum is defined simply as the product of the constant mass m_0 and the proper velocity:

$$p = m_0 v_{\text{pr}} \cdots \cdots (30)$$

An alternative interpretation of equation 29 arises if we combine the denominator with m_0 factor, and call this the *relativistic mass*, m, of the moving object, thus:

relativistic mass

$$m = \frac{m_0}{\sqrt{1 - v^2{}_{\text{lm}}/c^2}} \cdots \cdots (31)$$

In this approach momentum is defined as the product of this variable mass, m, and the improper velocity:

$$p = m v_{\text{lm}} \cdots \cdots (32)$$

It doesn't matter in the least which of these two definitions we choose for p. Experiments simply tell us the value of p; we may interpret this experimental value as we wish. The second viewpoint, summarized in equations 31 and 32, is perhaps the more fashionable, and it is the one we shall adopt in later Units. If one makes a series of measurements of the momentum of an object travelling at various speeds (v_{lm}), then, according to this second interpretation, equation 31, rewritten as $m = p/v_{\text{lm}}$, can be used to calculate the relativistic mass, m, at these different speeds. Figure 17 shows the results of such a study on the electron. (The momentum, p, is found by measuring the extent to which the electron's path is bent in a magnetic field.) Here the deduced values of m are expressed as fractions

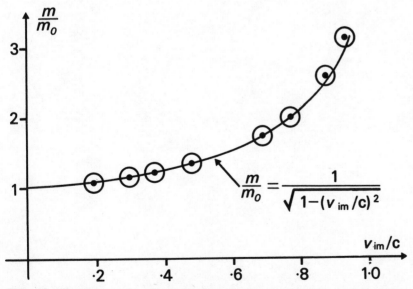

Figure 17 *Showing how the relativistic mass* m *of an electron varies with speed* v$_{\text{lm}}$. *Here the values of* m *are expressed as fractions of the rest-mass,* m$_0$, *while the speeds* v$_{\text{lm}}$ *are expressed as fractions of the speed of light,* c. *Superimposed on the experimental points is the variation to be expected from equation 31.*

45

of the rest mass, m_0, while the speeds, v_{im} are expressed as fractions of the speed of light, c. Superimposed on the experimental points is the variation of m/m_0 to be expected from equation 31. Of course, had we adopted the first interpretation of p, equation 30 rewritten as $v_{\text{pr}} = p/m_0$ would have told us how the proper velocity of the electron varied with the improper velocity. The experiments also demonstrate that p increases with v_{im}.

SAQ 18

A futuristic rocket ship with a rest mass of 10^6 kg is moving with an improper velocity of 0.9 c. (Take the velocity of light c as 3.0×10^8 m s^{-1}.) What is the momentum of the ship and its relativistic mass?

The detailed working of the problem is given on p. 61.

The momentum is 6.2×10^{14} kg m s^{-1}. The relativistic mass is 2.3×10^6 kg, which is 2.3 times greater than the rest-mass.

You should now do *SAQ*s 19–20, p. 56.

3.7 Recapitulation

We started off this Unit by making a study of some of the fundamental properties of electromagnetic waves; a study provoked by the desire to find the fastest possible method of transmitting information. We learnt that the velocity of light depends on the medium through which it travels, the velocity being greatest in a vacuum, and that the measured velocity of light is independent of the velocity of the light source and of the velocity of the observer. A logical consequence of these assumptions is that a clock mounted on a moving object will measure a shorter time, called the proper time, to cover a certain distance than will two clocks set out alongside the distance scale. The clocks alongside the scale measure the so-called improper time. However, we showed that the difference between the two timings can be ignored unless the speed of the object approaches that of light.

A systematic study of how the acceleration of an object depended on the accelerating force and on the mass of the object led to Newton's Second Law, formulated first in arbitrary units, then in SI units. It was possible to account for the behaviour of objects located in a rotating framework and to derive expressions for the centrifugal and centripetal forces. Newton's Second Law was also employed in the discussions of weight.

The problem of how to discuss the behaviour of an object which changes speed under the influence of hidden forces led to the formulation of momentum. Assuming Newton's Third Law it was proved that momentum should be conserved in collisions—a prediction borne out by experiment. However, for momentum to be conserved in collisions involving objects moving at speeds approaching that of light it must be given a more general formulation than mv. This general formulation was shown to predict that the mass of an object should increase as its speeds approach that of light, as is observed in practice.

Describing how things move

To most people, mention of the word 'velocity' or 'speed' would probably lead them to picture a car's speedometer. If asked what 54 m.p.h. meant they would probably, and correctly, reply that if the driver keeps going along so that the needle stays on the 54 m.p.h. mark he will cover 27 miles in $\frac{1}{2}$ an hour, 54 miles in 1 hour, 81 miles in $1\frac{1}{2}$ hours, etc. If asked how such a speedometer could be calibrated they might reply, although many are reluctant to invert the argument, that the car could be driven for, say, half an hour keeping the speedometer needle on some as yet unnamed point, and the total distance travelled could be measured. Had the car gone, say, 49 miles in 0.5 hour, then its speed was $49/0.5 = 98$ m.p.h. The reluctance to suggest such a technique may stem from knowing the problem of finding a free enough stretch of road where the car could be driven uniformly for anything like half an hour! But why not keep going for only $\frac{1}{4}$ hour (requiring a free stretch of $24\frac{1}{2}$ miles), or for 2 minutes or even 1, and measure the distances? The only possible objection is that it is difficult to measure these small distances, but cameras set up alongside the road to photograph the car's position at intervals of say 1 s, 0.1 s, or even 0.01 s will dispose of this objection (a scale set out along the road could be photographed simultaneously).

In fact the only way to be really certain of a car's speed is to measure the distance δs, covered in a short time δt (the symbol δ pronounced 'delta' just means 'a small amount of'). By division the speed of the car is

$$v = \frac{\delta s}{\delta t} \quad \dots\dots\dots\dots\dots\dots\dots(1)$$

The shorter the chosen value of δt, and hence the smaller the distance δs gone, the more precise the idea of velocity becomes.

Using equation 1, the car's speedometer can now be fully calibrated.

Suppose we go on a journey reading the car's speed at various times as we go along. A good way to present the information is to plot a graph of speed against time. Such a plot might look like that shown in Figure 18. On this particular journey the car started off in town, then travelled along a motorway, but had an accident. After it was repaired, the driver proceeded

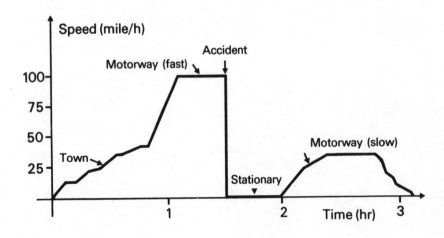

Figure 18 *Showing how the speed of a motor car might vary with time.*

more cautiously along the motorway. Finally he encountered some congested traffic. You should satisfy yourself that these seem reasonable interpretations of Figure 18.

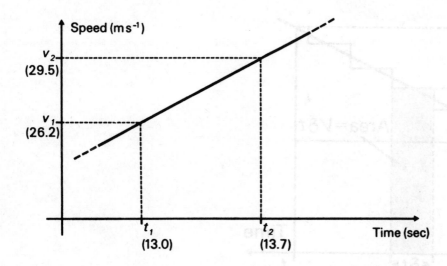

Figure 19 *Illustrating the particular case where the velocity of a car increases linearly with time.*

Rather than discuss this general case in more detail imagine that the plot was actually linear, as shown in Figure 19. Such a graph might possibly be obtained in practice if the driver kept his foot on the accelerator.

If you obtain a straight-line (i.e. a linear) graph, one of the first things you think about calculating is its slope. To determine the slope two lines may be drawn in, parallel to the velocity axis, starting at times t_1 and t_2. The corresponding velocities v_1 and v_2 may be read off from the graph. (Sample numerical values are included in Figure 19.) As the time changed from t_1 to t_2 (from 13.0 s to 13.7 s) the velocity changed from v_1 to v_2 (from 26.2 to 29.5 m s^{-1}). In other words the velocity changed by (v_2-v_1) ($=3.3$ m s^{-1}) in time (t_2-t_1) ($=0.7$ s). Therefore the velocity change per unit time, i.e. the acceleration, a, is given by

$$a=\frac{v_2-v_1}{t_2-t_1}$$

In the particular example given, $a=3.3$ m s$^{-1}/0.7$ s $=4.7$ m s^{-2}. You will see that a is the slope of the graph, which is constant throughout in a linear graph. Of course in the more general case shown in Figure 18 the slope, i.e. the acceleration, kept changing. For acceleration to have meaning in such a general case one must only examine the velocity change, δv occurring in a small time interval δt. Then

$$a=\frac{\delta v}{\delta t}$$

The smaller you make δt, the smaller become the possible wild fluctuations in δv, and the more meaningful becomes the acceleration a.

Another routine quantity to evaluate from graphs is the area underneath them. In the case of the graph showing constant acceleration we will evaluate the area under the straight line between time 0 and t. With this

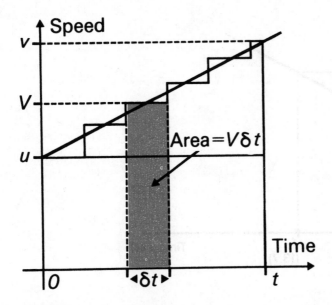

Figure 20   *Illustrating that the distance travelled by a car is the area under the graph of velocity plotted against time. We used the symbol V for the particular value of the velocity v at which we calculated the area Vδt under one step.*

end in view Figure 19 has been redrawn as in Figure 20; the original straight line has been replaced by a series of discrete steps. This stepped graph is constructed so that the area under it is the same as that under the original graph; the area under the latter is the sum of the areas under all the steps. (One is shown shaded.) Now the area under one of these steps is $v\delta t$, which is nothing more than *distance* gone in this time interval δt. Hence the total area between zero time and t represents the total distance gone in this period.

I.e.   in time t distance gone = Area under velocity-time curve between

time zero and t.

It is not hard to see that this area can be written in terms of the velocity at time zero (u, say) and that at time t (v, say). Thus

$$s = \text{Area} = \text{Area of rectangle} + \text{area of triangle}$$
$$= (ut) + \tfrac{1}{2}\,(\text{base} \times \text{height})*$$
$$= ut + \tfrac{1}{2}t\,(v - u)$$

i.e.
$$s = \frac{(u + v)t}{2} \quad \ldots\ldots\ldots\ldots\ldots (4)$$

* *See* MAFS, *section* 2.A.1.

This result can be put into different forms on recalling that the acceleration is the slope of the line (and constant).

I.e.
$$a = \frac{(v-u)}{t}$$

i.e.
$$at = v - u$$

or
$$v = u + at \quad \dots\dots\dots\dots\dots\dots (5)$$

Substituting equation 5 into equation 4 gives

$$s = \frac{(u + u + at)t}{2}$$

i.e.
$$s = ut + \tfrac{1}{2}at^2 \quad \dots\dots\dots\dots\dots (6)$$

An alternative expression, involving only v, u, and a, may be obtained by substituting the value of t given by equation 5, namely $t = (v-u)/a$, into equation 4.

$$s = \frac{(v+u)\,(v-u)}{2a}$$

$$s = \frac{v^2 - u^2}{2a} \quad \dots\dots\dots\dots\dots (7)$$

So now we have various alternative, but essentially identical, relations between the distances travelled in a time, t, by a body which was moving with a velocity u at time zero and accelerating with a constant acceleration a.

Where was the assumption made that the acceleration was constant?

The graph was supposed to be linear. The areas evaluated in equations 4, 6 and 7 are the areas under straight line graphs. Conversely if a body moves so as to obey equation 5, 6 or 7 its acceleration must be constant.

Appendix 2

Dimensional analysis applied to centripetal forces

Everyday experiences suggest that the inwardly directed force F_1 required to keep an object rotating in a circular path at, say, the end of a spring depends on the mass m of the object, the radius r of its path and the speed v at which it is moving. Other factors such as, for example, the shape of the object are of no obvious importance. If it should not prove possible to make the dimensions of F_1 agree with those of the product of m, v, and r, each raised to an appropriate power, then other factors will have to be considered. Even if agreement is obtained, it is still possible that other factors have been neglected—dimensional analysis may not yield a unique solution. But assuming

$$F_1 = \text{constant} \times m^\alpha v^\beta r^\gamma$$

Putting in dimensions of force, mass, velocity and radius, gives

$$MLT^{-2} = M^\alpha \left[\frac{L}{T}\right]^\beta L^\gamma$$

—force, remember, has the dimensions of mass × acceleration (Newton's Second Law).

Equating the powers of M, L, and T on the right-hand side with those on the left-hand side gives:

$$\text{for mass } \alpha = 1 \quad \dots \dots \dots \dots \dots \dots (1)$$

$$\text{for length } 1 = \beta + \gamma \quad \dots \dots \dots \dots \dots (2)$$

$$\text{for time } -2 = -\beta \quad \dots \dots \dots \dots \dots (3)$$

$$\therefore \quad F = \text{constant} \times \frac{mv^2}{r}$$

We have now combined m, v, and r in the only way that the units of the combination are those of force. Dimensional analysis is, of course, powerless to indicate what the dimensionless constant might be.

Appendix 3 (Black)

Momentum—a more general discussion

Newton's Second Law, equation 15, relates the acceleration a produced in a body of mass m by a force F, viz.

$$F = ma$$

But acceleration, being the rate of change of velocity can be written as

$$a = \frac{dv}{dt}$$

$$\therefore \quad F = m\frac{dv}{dt} \text{ or, provided } m \text{ is a constant,}$$

$$F = \frac{d}{dt}(mv)$$

or

$$F = \frac{dp}{dt} \quad \dots\dots\dots\dots\dots(1)$$

where the substitution $p = mv$ has been made, p being the momentum of the body.

Cross-multiplying equation (1) gives

$$F dt = dp \quad \dots\dots\dots\dots\dots(2)$$

Integrating equation (2) gives

$$\int F dt = \int dp \quad \dots\dots\dots\dots\dots(3)$$

If a force acts during the interval t_1 to t_2 as a result of which the momentum of the body changes from p_1 to p_2 then

$$\int_{t_1}^{t_2} F dt = \int_{p_1}^{p_2} dp$$

$$\int_{t_1}^{t_2} F dt = \left[p \right]_{p_1}^{p_2}$$

$$\therefore \quad \int_{t_1}^{t_2} F dt = p_2 - p_1 \dots\dots\dots\dots\dots(4)$$

The left-hand side of equation 4 is the area under a graph showing how the force acting on the body changes with time, between times t_1 and t_2. This so-called impulse of the force is, as equation 4 demonstrates, equal to the resulting change in the momentum of the body.

Self-Assessment Questions

Section 3.2

Question 1 (*Objective 2*)

Which, if any, of the following properties of light are assumed in formulating the theory of special relativity?

1 The velocity of light exceeds that of sound.
2 The wavelength of blue light is less than that of red light.
3 The velocity of light does not depend on the speed of its source.
4 No optical experiments performed within a laboratory have ever shown up any uniform motion of the laboratory.

Section 3.3

Question 2 (*Objective 3*)

The experiment described in section 3.3.1 is performed in a car with $L = 1.5$ m moving at a speed of 2.0×10^8 m s^{-1}. Calculate the proper and improper times for the light to traverse the path from the bulb to the mirror and back to the clocks.

Question 3 (*Objective 3*)

In 1965 Ron Clarke ran the 10 000 metre event in 27 mins. 39.4 secs. Had Clarke carried a watch, how would his timing have differed from the judges? Needless to say the judges were stationary alongside the track.

Question 4 (*Objective 3*)

The proper velocity of an object:

1 Is greater than the improper velocity of the object.
2 Is less than the improper velocity of the object.
3 Is equal to improper velocity of the object.
4 Is equal to the speed of light in a vacuum.

Section 3.4

Question 5 (*Objective 5*)

In an inertial frame:

1 Newton's First Law holds true but optical experiments enable one to detect that the frame in which the experiments are performed is moving.
2 Newton's First Law holds true.
3 Everything behaves in a sluggish fashion.
4 Newton's First Law holds true, provided that the speed is much less than the speed of light.

Section 3.5

Question 6 (*Objective 4*)

A car accelerates away from rest at 4 m s^{-2}. What is its velocity after it has gone 8 m?

Question 7 (*Objective 4*)

A sprinter who is travelling at 2 m s^{-1} accelerates at a constant rate while he travels 8 m. His final velocity is 5 m s^{-1}. How long does he spend accelerating?

Question 8 (*Objective 4*)

A puck accelerates from rest at a constant rate for 3 s through a distance of 5 m. What is the puck's acceleration?

Question 9 (*Objective 5*)

Make an order of magnitude estimate of the push that human legs can provide while accelerating away from rest.

Question 10 (*Objective 5*)

A car of mass 400 kg accelerates away at a constant rate from rest. In 15 s it has reached a speed of 50 km per hour. What force is the engine providing?

Question 11 (*Objective 5*)

An ascending escalator set at 45° to floor level is moving at 0.8 m s^{-1}. Someone is walking up the escalator at 1.1 m s^{-1}. Use the head-to-tail rule to deduce his speed and direction relative to someone stationary at the foot of the escalator.

Question 12 (*Objective 5*)

What roughly is the force of attraction between the Earth and a 'quarter pound' slab of chocolate? You should know that, when released, a slab falls with a constant acceleration. In an actual experiment a slab in falling from rest through a distance of 3 m acquired a final speed of 7.5 m s^{-1}.

Question 13 (*Objective 5*)

Calculate the force necessary to keep a 500 gm bag of sugar (about 1 lb.) revolving in a circle of radius 0.75 m with a period of 0.4 s.

Question 14 (*Objectives 5*)

In the SI system of units:
1 Experiment shows that the force required to accelerate a mass of 1 kg at a constant rate of 1 m s^{-2} is 1 N.
2 The weight of 1 kg is 1 N.
3 Densities are measured in kg m^{-3}.
4 Centripetal forces are measured in kg m s^{-1}.

Question 15 (*Objective 5*)

An inwardly directed force is required to keep an object moving round in a circular path because:

1 The velocity of the object keeps changing.
2 The speed of the object keeps changing.
3 The acceleration of the object keeps changing.
4 The mass of the object depends on its orientation in space.

Question 16 (*Objective 5*)

The angular frequency, ω, of an object traversing a circular path of radius r in a periodic time T is defined as:

1 $2\pi r/T$.
2 $2\pi/T$.
3 $2\pi/rT$.
4 $2\pi T/r$.

Section 3.6

Question 17 (*Objective 6*)

A 90 kg person is a passenger in a car travelling at 60 km hour⁻¹, which has a head-on collision and stops 'dead'. If the car seat-belt holding the passenger 'gives' for 0.2 s calculate the force which the seat-belt exerts on the passenger as he is brought to rest. Assume the belt provides a constant force.

Question 18 (*Objective 6*)

A futuristic rocket ship with a rest mass of 10^6 kg is moving with an improper velocity of 0.9 c. (Take the velocity of light c as 3.0×10^8 m s⁻¹.) What is the momentum of the ship and its relativistic mass?

Question 19 (*Objective 6*)

In a collision between two objects:

1 the total momentum before and after the collision is always the same;
2 the total momentum before and after the collision is never the same;
3 the total momentum before and after the collision is sometimes the same;
4 momentum is not conserved but the rate of change of momentum is conserved.

Question 20 (*Objective 6*)

As the speed of an object increases towards that of light, its relativistic mass:

1 Falls towards zero.
2 Increases towards infinity.
3 Remains constant.
4 Oscillates sinusoidally.

Self-Assessment Answers and Comments

Question 1

Answers 3 and 4 are correct. See section 3.2.3.

Question 2

The proper time t_{pr} is the time the light flash takes to travel to the mirror and back (a distance of $1.5+1.5=3.0$ m) as judged by a passenger in the car. Since to him light travels at 3.0×10^8 m s^{-1}, the time taken is

$$t_{pr} = \frac{\text{distance}}{\text{velocity}}$$

$$= \frac{3.0 \text{ m}}{3.0 \times 10^8 \text{ m s}^{-1}}$$

$$\therefore \quad t_{pr} = 1.0 \times 10^{-8} \text{ s.}$$

During the time $\frac{1}{2}t_{pr}$ the car travels a distance as measured by an observer on the road of (the car's speed) $\times \frac{1}{2}t_{pr} = 2.0 \times 10^8 \times 0.5 \times 10^{-8}$ s $= 1$m. This corresponds to the halfway stage shown in Figure 2. This calculation assumes that the car's speed that is specified is 'the distance gone as measured alongside the road divided by the time for the journey as measured on the car'. The distance travelled by the flash as measured by roadside observer during the interval is therefore $\sqrt{1^2+1.5^2}$ (in terms of Fig. 2 it is, applying Pythagoras' theorem $= \sqrt{L^2+(l/2)^2}$) which is $\sqrt{3.25}=1.8$ m. Therefore the total path length of the flash as seen by the ground-based observer is $2 \times 1.8 = 3.6$ m. Since light also travels at 3.0×10^8 m s^{-1}, to this observer the improper time, t_{im}, is

$$t_{im} = \frac{3.6 \text{ m}}{3.0 \times 10^8 \text{ m s}^{-1}}$$

$$t_{im} = 1.2 \times 10^{-8} \text{ s.}$$

Question 3

Equation 6 shows that:

$$t_{pr} = t_{im}\sqrt{1-v^2_{im}/c^2}$$

Question 1 pinpointed the reason why t_{pr} is shorter than t_{im}.

Here t_{im}, the timing of the stationary judges who are separated by a distance of 10^4 m, is 27 mins. 39.4 secs. $=(27 \times 60)+39.4$ s$=1\,659.4$ s,

i.e. $\qquad\qquad\qquad t_{im} = 1\,659.4$ s.

In the judges' opinion Clarke ran at a speed of v_{im}, given by

$$v_{im} = \frac{\text{distance measured by stationary judges}}{\text{judges' timing}}$$

$$= \frac{10^4 \text{ m}}{1\,659.4 \text{ s}}$$

$$\therefore \quad v_{im} = 6.035 \text{ m s}^{-1}.$$

Substituting $v_{im} = 6.035$ m s^{-1} and $c = 3.0 \times 10^8$ m s^{-1} into equation 6 gives

$$t_{pr} = t_{im}\sqrt{1-(6.0)^2/(3 \times 10^8)^2}$$

(It is quite adequate to take $v_{im}=6.0$ m s^{-1}, because the second term under the root sign is so much smaller than unity.)

i.e. $\qquad\qquad t_{pr} = t_{im}[(1-(6.0)^2/(3 \times 10^8)^2)^{\frac{1}{2}}]$

57

According to the binomial theorem (equation 3) when x is very much less than unity

$$(1+x)^m \approx 1 + mx + \text{second-order terms.}$$

Here $x = -(6.0)^2/(3 \times 10^8)^2$ and $m = \frac{1}{2}$.

So, to a very good approximation

$$t_{pr} \approx t_{lm}[1 - \frac{1}{2}((6.0)^2/(3 \times 10^8)^2)]$$

$$\approx t_{lm}\left(1 - \frac{36}{18 \times 10^{16}}\right)$$

$$\approx t_{lm}\left(1 - \frac{2}{10^{16}}\right)$$

In other words t_{pr}, Clarke's own estimate of the timing would be less than that of the judges by 2 parts in 10^{16}.

Question 4

Answer is 1. Note that t_{pr} is always less than t_{lm} unless $v_{lm} = 0$. Refer back to equation 6 in section 3.3.2.

Question 5

Answer is 2. See section 3.4.

Question 6

The information given is the initial speed u ($=0$), the acceleration a ($=4$ m s^{-2}) and the distance gone s ($=8$ m). What is required is the final velocity v.

Equation 7 of the Appendix 1 shows that

$$v^2 - u^2 = 2as$$

or

$$v^2 = 2as + u^2, \text{ or substituting for } u, a,$$
$$\text{and } s$$

$$v^2 = (2 \times 4 \times 8) + 0 = 64 \text{ m}^2 \text{ s}^{-2}$$

$$\therefore \quad v = 8 \text{ m s}^{-1}.$$

If you do not know how equation 7 was arrived at, read pages 49–51 again.

Question 7

Here $u = 2$ m s^{-1}, $v = 5$ m s^{-1}, and $s = 8$ m. The problem is to find the time spent accelerating at a constant rate.

Equation 4 of Appendix 1 is relevant:

$$s = \left(\frac{u+v}{2}\right)t$$

$$\therefore \quad t = \frac{2s}{u+v} = \frac{16 \text{ m}}{(2+5) \text{ m s}^{-1}}$$

$$= \frac{16}{7}\text{s}$$

$$\therefore \quad t = 2.3 \text{ s}.$$

A useful test of your understanding of equation 4 is to ask yourself where the factor of 2 originates.

Question 8

Here $u=0$, $t=3$ s and $s=5$ m. The problem is to find the acceleration a.

Equation 6 of the Appendix 1 is the relevant relation.

i.e. $s=ut+\frac{1}{2}at^2$

or $a=\dfrac{2(s-ut)}{t^2}$

Substituting for u, s and t gives:

$$a=\frac{2(5-0)}{3^2}\frac{m}{s^2}$$

$$=\frac{10}{9}\text{ m s}^{-2}$$

$$\therefore\quad a=1.1\text{ m s}^{-2}$$

Question 9

An average sort of mass to assume for an adult is around 10 stone, i.e. 140 lbs, i.e. $140/2.2 \approx 60$ kg; let us say 10^2kg. One might be able to sprint up to a speed of about the same as that kept up by a four minute miler! i.e. a top speed of around $1500m/4 \times 60s \approx 6m\text{ s}^{-1}$ (1 mile $\approx$ 1500m). Conceivably it would take 2 s to 6 s to reach this speed, i.e. one's acceleration is of order $6\text{ m s}^{-1}/4\text{ s} \approx 1\text{ m s}^{-2}$. Applying equation 14 gives as the force F required to produce this acceleration, a, in a body of mass m of 10^2 kg as

$$F=ma$$

$$=10^2 \times 1\text{ kg m s}^{-2}$$

$$F=10^2\text{ N}$$

Since a leg muscle might well be able to lift up to 10^2 bars of chocolate, each of weight 1 N, it seems a plausible enough answer.

Question 10

The car's constant acceleration must first be determined from the information that it started from rest, $(u=0)$ and after a time t of 15 s it had reached speed u of 50 km per hour. The defining relation for acceleration is $a=(v-u)/t$ but you get an answer in mixed units involving hours and seconds if you simply substitute $u=0$, $v=50$ km per hour, and $t=15$ s. We should convert 50 km per hour into units of m s^{-1}.

$$\frac{50\text{ km}}{1\text{ hour}}=\frac{50\times10^3m}{1\times60\times60s}$$

$$=\frac{50\times10^3}{3.6\times10^3}\frac{m}{s}$$

$$=\frac{50}{3.6}\text{ m s}^{-1}$$

$$\therefore\quad a=\frac{50}{3.6\times15}\frac{ms^{-1}}{s}=0.925\text{ m s}^{-2}$$

Applying Newton's Second Law, equation 14 gives the accelerating force provided by the engine as $400\text{ kg}\times0.925\text{ m s}^{-2}=370\text{ kg m s}^{-2}=370\text{ N}$.

Question 11

The individual vectors to be added are shown in Figure 21, where lines have been drawn of length proportional to 0.8 m s⁻¹ and 1.1 m s⁻¹ respectively in the required directions. Putting the head of one vector at the tail of the other and then joining the first head to the last tail as in the figure gives the resultant. To a ground-based observer, the person on the escalator is moving up at 1.9 m s⁻¹ at 45° to the ground. If you are still uncertain about how vectors should be added, read *MAFS*, section 4.D.

Question 12

First the acceleration of the bar of chocolate to the ground must be calculated. As we know the mass ($\frac{1}{4}$ lb is approximately $\frac{1}{4} \times \frac{1}{2.2}$ kg $= 10^{-1}$ kg) this will enable the force acting on the bar to be found. We are given $u = 0$, $s = 3$ m $v = 7.5$ m s⁻¹. We want a.

$$\therefore \quad a = \frac{v^2 - u^2}{2s}$$

$$\text{i.e. } a = \frac{56 - 0}{6} = 9.3 \text{ m s}^{-2}$$

$\therefore$ The accelerating force F is, from equation 14, given by

$$F \approx 10^{-1} \times 9.3 \text{ kg m s}^{-2}$$

$$F \approx 1.0 \text{ N}$$

Question 13

The inwardly directed force, F_1, necessary to keep a body of mass m moving in a circle of radius r with a periodic time T s, i.e. with a speed $2\pi r / T$, is given by

$$F_1 = mv^2/r$$

$$= m\left(\frac{2\pi r}{T}\right)^2 \Big/ r$$

$$= \frac{4\pi^2 r m}{T^2}$$

Here $r = 0.75$, $m = 0.5$ kg, $T = 0.4$ s

$$\therefore \quad F_1 = \frac{4\pi^2 \times 0.75 \times 0.5}{0.4^2} \frac{m \text{ kg}}{s^2}$$

$$= \frac{39.5 \times 0.75 \times 0.5 \text{ N}}{0.16}$$

$$= 92.5 \text{ N}$$

Question 14

Answer is 3. See section 3.5.2. Answer 1 is wrong. The newton is defined to be the force which gives 1 kg an acceleration of 1 m s⁻².

Question 15

Answer is 1. See section 3.5.5. Remember the difference between speed and velocity.

Question 16

Answer is 2. See section 3.5.5.

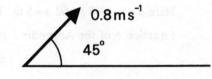

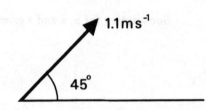

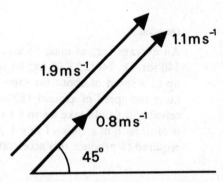

Figure 21 *Showing how two vectors representing velocities of 0.8 m s⁻¹ and 1.1 m s⁻¹ are combined by the 'head-to-tail' rule to give a resultant of 1.9 m s⁻¹.*

Question 17

The momentum p of the passenger in the moving car is given by,

$$p = mv$$

$$= 90 \times 60 \times 10^3/3600 \text{ kg m s}^{-1}$$

$$= 1500 \text{ kg m s}^{-1}$$

Since the passenger ends up with zero velocity p also gives the change in his momentum. In bringing the passenger to rest the belt exerts a supposedly constant force F for a time t, such that the impulse Ft equals the change in momentum p (equation 26)

$$\text{i.e. } Ft = p$$

$$\text{i.e. } F = \frac{p}{t}$$

$$= \frac{1500 \text{ kg m s}^{-1}}{0.2 \text{ s}}$$

$$\therefore \quad F = 7.5 \times 10^3 \text{ N.}$$

Figure 13 may help remind you why impulse equals change in momentum.

Question 18

From equation 31, the relativistic mass is

$$m = \frac{m_o}{\sqrt{1 - v_{im}^2/c^2}}$$

Substituting $m_o = 10^6$ kg, $v_{im} = 0.9c$, gives

$$m = \frac{10_6}{\sqrt{1 - \frac{(0.9c)^2}{c^2}}} \text{ kg}$$

$$= \frac{10^6}{\sqrt{0.19}} \text{ kg}$$

$$= 2.3 \times 10^6 \text{ kg}$$

So, from equation 32, the momentum is

$$p = mv_{im}$$

$$= 2.3 \times 10^6 \times 0.9 \times 3 \times 10^{-8} \text{ kg m s}^{-1}$$

$$= 6.2 \times 10^{14} \text{ kg m s}^{-1}$$

Question 19

Answer is 1. See section 3.6.2.

Question 20

Answer is 2. See section 3.6.3.

Notes

The Open University

Science Foundation Course Unit 4

FORCES, FIELDS AND ENERGY

Prepared by the Science Foundation Course Team

THE OPEN UNIVERSITY PRESS

Contents

Table A

A List of Scientific Terms, Concepts and Principles used in Unit 4

Taken as pre-requisites			Introduced in this Unit			
1 Assumed from general knowledge	**2** Introduced in a previous Unit	Unit No.	**3** Developed in this Unit	Page No.	**4** Developed in a later Unit	Unit No.
operation	weight	3	torsion balance	11	proton	6
	newton balance	3	law of gravitational force	12	neutron	6
magnet	cosmic rays	2	cell	13	electron	6
	field	2	conductor	14	nucleus	6
	centrifugal force	3	insulator	14	elementary particle	32
	inertial frame	3	coulomb	15	strong interaction	31
	positive charge	2	current	15	weak interaction	32
	negative charge	2	amp	16		
	relativistic mass	3	ampere	16		
	rest mass	3	ammeter	17		
	proper time	3	current balance	17		
	improper time	3	electrodynamic force	17		
	newton	3	electrostatic force	17		
	electron	2	dielectric constant	19		
	muon	2	relative permittivity	19		
	Newton's Second Law	3	electric field	22		
			gravitational field	24		
			energy	25		
			thermal energy	25		
			Calorie	27		
			calorie	27		
			joule	27		
			potential energy	28		
			electrical potential	30		
			potential difference	30		
			volt	30		
			electron volt	31		
			kinetic energy	32		
			power	32		
			watt	32		
			horse-power	33		
			relativistic energy	34		
			rest-mass	37		
			rest-mass energy	37		
			total relativistic energy	37		

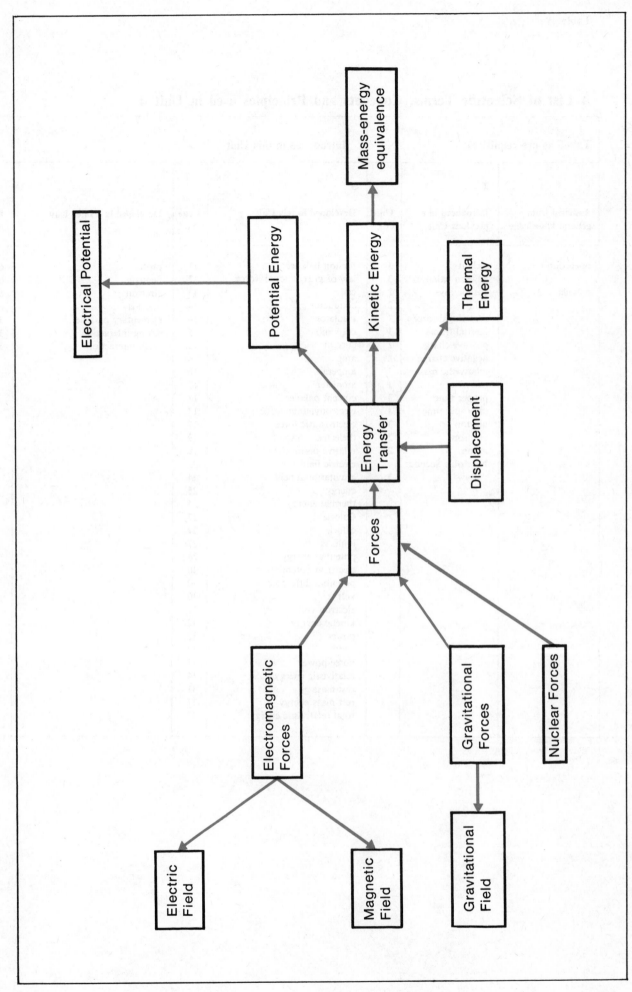

Objectives

After you have completed your study of this Unit, you should be able to:

1 Define correctly, or recognize the best definition of all the terms, concepts and principles, in column 3 of Table A.

2 (Tested in *SAQ* 1)
State how the force of gravitational attraction between two spheres depends on the mass of the spheres and on their separation and be able to make simple calculations of the magnitude of these forces.

3 (Tested in *SAQ* 2)
State the principal factors which affect the force between two current-carrying wires.

4 (Tested in *SAQ* 3)
State how the electrostatic force between two charged spheres depends on the magnitude of the charges and on the separation of the spheres and be able to perform simple calculations of the magnitude of these forces.

5 (Tested in *SAQ* 4, 5)
Perform simple calculations which convert electrical currents measured in amps and electrical charges measured in coulombs into the equivalent numbers of electronic charges.

6 (Tested in *SAQ* 6, 7)
Perform simple calculations of the magnitudes of electric fields.

7 (Tested in *SAQ* 8)
State how the energy transferred in mechanical processes is related to the force, to the distance through which the force acts, and to the angle which the force makes with displacement, and to be able to perform simple calculations of such energy transfers.

8 (Tested in *SAQ* 9, 10, 12)
Perform simple calculations of kinetic and potential energies including calculations of electrical potential energies.

9 (Tested in *SAQ* 11)
State how the power of an electrical device is related to the potential difference across the device and the current flowing through it.

4.1 Introduction

In the previous Unit you learnt that, in an inertial frame of reference, an object left alone remains at rest (Newton's First Law). You then learnt of the effect of applying forces of various magnitude to bodies of different masses; a study culminating in the formulation of Newton's Second Law. Throughout these experimental studies, the forces came either from human pushes and pulls or from wire springs. Such forces as these are hardly in any sense 'basic'; they are undoubtedly the macroscopic expression of other, more fundamental forces. You might think that to explain all the different forces with which you are familiar—ranging from muscular pushes, through the workings of a petrol engine, to what happens in nuclear explosions—it would be necessary to postulate many such fundamental forces. Surprisingly, it turns out that the entire spectrum of macroscopic forces can be explained in terms of only four basic forces: gravitational, electromagnetic and the strong and weak nuclear inter-actions. In this Unit we shall begin by outlining some of the main charac-teristics of the first three of these four forces (later Units will take up the story of how these forces explain at least some of the everyday pushes and pulls). This will lead us to look more closely at what happens when forces act and objects are made to move. We shall begin to formulate a concept of energy.

4.2 The Fundamental Forces

4.2.1 Gravitational forces

If there is one force of which we are constantly being reminded, it is the pull of the Earth on ourselves, our weight. Expressed more objectively, what we are saying is that there is an attractive force between a large lump of matter (the Earth), and a smaller lump (ourselves). But do *any* two lumps of matter attract each other?

We all know that gravitational attraction does not, for example, cause the potatoes in a polythene pack to adhere together and that even massive objects like buildings or mountains exert no detectable physical pulls on ourselves. Were such pulls of the magnitude of a newton we would, of course, be very conscious of their presence. However, before concluding that, aside from our weight, gravitational forces are undetectable in everyday situations, it might be wise to perform a few experiments where the forces are very consciously sought.

The following experiments may sound like poor party games, but they are perfectly respectable scientific investigations. However, if you are 'certain' you know the conclusions in advance, there may be little point in performing these particular experiments; instead encourage some totally unsuspecting person to carry them out.

> **Experiment 1**
> **The aim of this experiment is to detect gravitational forces. Hold something massive in each hand. Bring the objects closer together and try to decide if they attract each other. You should use a variety of objects, such as, for example, books, rocks, bags of potatoes, even cabbages; the effect may well depend on what the objects are made of. After all, our chemical composition and that of the Earth are very different. Another worthwhile variant is to close your eyes and ask someone else to bring these objects up to your outstretched hand. You, by sensing any pull on your hand, may be able to tell if the object is there.**

So, although the Earth pulls on us, there is no gravitational attraction between two much less massive pieces of matter? No, it could simply be that the forces between our objects are less than the forces we can detect.

> **Experiment 2**
> **What is the smallest force your hands can detect? One way to find out is to get an average apple which will weigh about 1 N and slice it up into progressively smaller pieces (counting the divisions as you go) until, with your eyes closed, you just cannot tell if a piece is on the palm of your hands. (Try shaking your palm around.) If you can sense, for example, the weight of a one hundredth part of an apple then you will be sensing a force of 10^{-2} N. Another technique is to use a pad of paper whose weight you guess, to an order of magnitude, prior to tearing it up into successively smaller pieces. Of course the experiment is a crude one, but we are only after an order of magnitude value of the minimum force you can respond to.**

You may well be able to detect forces of 10^{-4} N to 10^{-6} N, and yet have failed to detect any evidence for gravitational attraction between your experimental objects. A more sensitive force-measuring device than ourselves is clearly required.

One such device, known as a *torsion balance*, is shown in Figure 1.
Basically it consists of a fine quartz wire, W, carrying a light horizontal
beam, B. A small force applied at either end of the beam will cause the
wire to twist in much the same way as a gentle breeze can blow a swing
door open. Some such balances can be made to respond to forces as
small as 10^{-12} N. As you can see in Figure 1, two lumps of matter, of
equal mass m_1, are hung from the ends of the beam; if they move, the
beam will move. Two fixed lumps of matter, of equal masses, m_2, are
then placed on opposite sides of the beam, each at a distance r from the
moveable masses. Any attractive forces between the fixed and the moveable masses will cause the beam to rotate and the wire to twist. And the
wire does twist; so gravitational forces do exist, aside from the Earth.

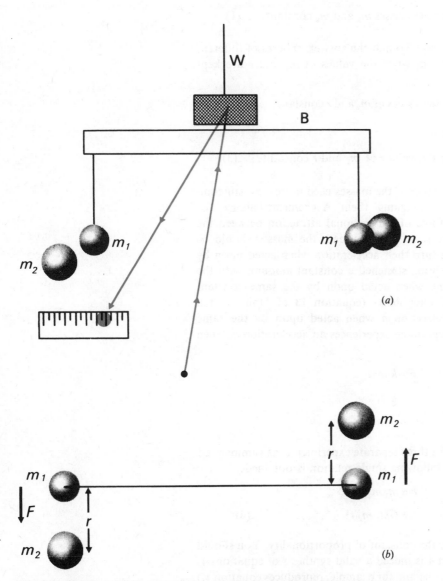

Figure 1 *Gravitational forces studied by means of a torsion balance.* (a) *A light horizontal
beam, B, carrying the masses* m_1, *is hung from a fine quartz wire W. A light beam reflected
off the mirror onto a scale enables the rotation of the beam to be measured.* (b) *A plan of
the apparatus.*

The obvious next question is how does the force of attraction depend
on the separation of the masses. But, should the objects happen to be,
say, a bag of potatoes and a bunch of bananas, how will this separation
be measured?

**What shape objects should we use to make it easy to specify how far they are
apart?**

11

Uniform spheres are symmetrical and have well-defined centres and so are preferable to objects like bunches of bananas. Therefore spheres will be employed, which should be preferably of a high density material like lead or gold so that there is plenty of matter around. There is plenty of matter in the Earth and it pulls hard on us.

To study how the force of attraction F between the fixed and the moveable spheres of mass m_1 and m_2 is related to their separation r, one need simply study how the angle of twist of the wire (normally measured by means of a light beam reflected off a mirror that is mounted on the wire) varies with the positioning of the fixed spheres. The result is that, when r is doubled, F, as deduced from the calibrated twist, goes down by a factor of four. When r is trebled, F is reduced ninefold. In general,

$$F \propto 1/r^2 \text{ keeping the masses } m_1 \text{ and } m_2 \text{ constant} \ldots (1)$$

How does F depend on, say, m_1? To find the answer, spheres of different masses are hung from the beam while the values of m_2 and r are kept constant. The result:

$$F \propto m_1 \text{ keeping the values of } m_2 \text{ and } r \text{ constant} \ldots (2)$$

When m_2 is varied the result is:

$$F \propto m_2 \text{ keeping the values of } m_1 \text{ and } r \text{ constant} \ldots (3)*$$

As a matter of principle the values of the masses used in this investigation should not be determined by 'weighing' them. A chemical balance, for example, makes use of the force of gravitational attraction between the Earth and the objects in the scale pans. Instead, the masses should be determined by comparing in turn their acceleration when acted upon by a constant force, such as a spring stretched a constant amount, with the acceleration of a kilogramme when acted upon by the same constant force. As Newton's Second Law shows (equation 13 of Unit 3), if a mass m experiences an acceleration a when acted upon by the same constant force F and the kilogramme experiences an acceleration a_1 when acted upon by F then

$$F = k\, m\, a$$
$$= k\, a_1$$

or
$$m = a_1/a.$$

Gathering up the results of the three separate experiments, as summarized in equations 1, 2 and 3, the following single relation is obtained:

$$F \propto m_1\, m_2/r^2$$

or
$$F = Gm_1\, m_2/r^2 \ldots\ldots\ldots\ldots (4)$$

where G has been written for the constant of proportionality. You should satisfy yourself that equation 4 is indeed a valid synthesis of equations 1, 2 and 3. (Keeping m_1 and m_2 constant, for example, reproduces equation 1.) Equation 4 cannot, of course, be expected to describe how, say, two cylinders interact. Indeed the $1/r^2$ dependence (equation 1), the so-called inverse square law relation, would not have been found had cylinders or any other shaped objects been employed. The simplicity of the end result is, in itself, a valid enough reason for employing spheres.

---

* You might well have been able to predict that equation 3 should have the same form as equation 2. Were the two equations different, the measured twist would depend on which mass was hung from the beam and which was fixed—Nature would be 'lopsided'. As you will learn from later Units, much modern research in nuclear physics is in fact devoted to searching for evidence of 'lopsidedness' in Nature, or its absence.

As the experiments have been carried out with known masses at known separations, they produce the value of G directly. For example, when a lead sphere of mass 7.4 kg is placed 7.0×10^{-2} m from a gold one of mass 2.6×10^{-3} kg the force of attraction is 2.62×10^{-10} N.

Using this data, deduce the value of G for yourself.

Rearranging equation 4 gives

$$G = Fr^2/m_1\, m_2$$

$$= (2.62 \times 10^{-10}) \times (7.0 \times 10^{-2})^2/7.4 \times 2.6 \times 10^{-3} \quad \text{N m}^2 \text{ kg}^{-2}$$

$$= 2.62 \times 49.0 \times 10^{-14}/7.4 \times 2.6 \times 10^{-3} \quad\quad\quad \text{N m}^2 \text{ kg}^{-2}$$

So $\quad G = 6.67 \times 10^{-11}$ N m^2 kg^{-2}.

Since 1 N = 1 kg m s^{-2} (Unit 3, p. 30), this result may be written as $G = 6.67 \times 10^{-11}$ m^3 kg^{-1} s^{-2}. (If these manipulations of units worry you read section 3 of the handbook *The Handling of Experimental Data*.) What is interesting is not the particular numerical value of G but the fact that, within the limits of experimental error, G is constant for a wide range of materials. This constancy suggests that we are indeed dealing with something basic.

SAQ 1
Make an order-of-magnitude estimate of the force of gravitational attraction between two adult human beings when standing side by side. Using a relation derived for spheres is bound to make the answer suspect; guesses of human masses correct to a factor of two or so will therefore be quite adequate. The problem is worked out on p. 45.

The force is about 10^{-5} N.

Compare this force with the minimum force that *you* could detect in experiment 2.

Off-hand it looks as if we might *just* be able to detect the presence of objects of about human mass when we are close to them—or to detect larger objects which are further away. It is not hard to prove that similar magnitude pulls can be expected in mountainous country! Indeed an early method of determining G was to measure the angle which a plumb line (carrying a spherical bob) made with respect to the vertical when it was located close to mountains. In a later Unit (22) you will learn how a detailed study of the Earth's gravitational pull can provide important clues about the Earth's structure.

4.2.2 Electromagnetic forces

Early in the nineteenth century, it was discovered that, if one dipped metal plates into certain solutions, interesting things happened when the plates of these *cells*, or batteries, were connected together in various ways. The sort of circuit which was employed is shown in Figure 2 (a) where the batteries are indicated by pairs of lines of unequal length; unequal to

cell (battery)

emphasize that the two plates have different compositions. The circuits are completed by lengths of metallic wire which run reasonably close to each other. Switches are incorporated as a convenient way of making and breaking the circuit. We may summarize the experiments (some of which you will see being performed in the TV programme of this Unit) and their findings, as follows:

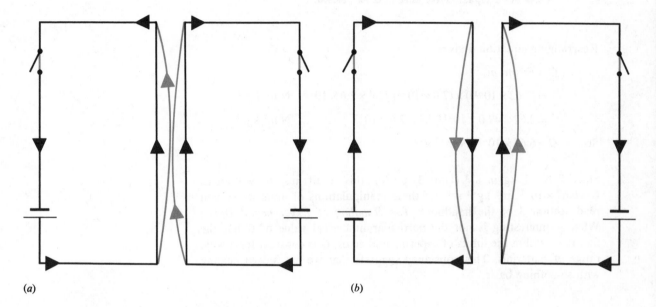

(a)   (b)

Figure 2 Electrodynamic forces. Depending on the sense in which the batteries are connected up, the force between the wires may be attractive, as in (a), or repulsive, as in (b).

1   When both switches are closed, each wire is attracted towards the other. The force is only present when *both* switches are closed; open one and the wires will cease to attract. Since the masses of the wires do not change when the switches are closed, the force can hardly be gravitational in origin. So some other force must be responsible.

2   When the connections to one of the batteries are reversed as in Figure 2 (b), and the first experiment is repeated, the wires, instead of being attracted together, are now pushed apart. Of course if the connections to the second battery are also reversed, the circuit will be back to that of Figure 2 (a) (turned upside down), so this further experiment need not be performed. (Strictly speaking, however, this is just the sort of experiment that *should* be performed, *a priori*, there is no reason why the experiment might not behave differently when 'upside down'. In this case, however, it behaves the same.)

3   When the number of batteries in the circuits is increased, the forces of attraction or repulsion are also increased.

4   When the wires are moved further apart the forces decrease. Changing the angle between the wires also changes the forces.

5   When the experiments are performed with the circuits connected up by lengths of plastic, nothing moves, i.e. no forces are present. Materials like plastics which give no effect are called *insulators*; those that do, *conductors*.

insulators and conductors

6   When the medium in which the wires are located is changed the forces may change. For example, lumps of iron placed between the wires will increase the forces.

14

Apart from heating up, the wires do not change in any obvious way when the switches are closed. However, there is a simple demonstration which does suggest that the internal state of the wires does change. As you will see in the TV programme, the evidence suggests that something flows through conductors that are connected across electrical supplies, such as batteries. One can therefore, as in Figure 2, put in arrows indicating the direction of flow of this 'something'*.

The Broadcast Notes contain further details of the experiment. Unfortunately this experiment gives no definitive answer as to the nature of this 'something'. It does not tell us, for example, whether it is something discrete (like a stream of particles) or something continuous (like a so-called 'classical' fluid: one with no obvious atomic character). Whatever the nature of this 'something', it would be very convenient to be able to specify the magnitude of the *current*, in other words how much of this 'something' is flowing past a fixed point in the wire per second. Indeed, the conventional way of speaking of an electrical current as so many *coulombs of charge* flowing past a point per second is very convenient, since we could, if we wished, imagine 'coulombs of charge' to stand for either 'number of particles' or 'cubic metres of fluid', depending on how we picture the 'something'.

current

coulomb

To frame an operational definition of unit current one must, of course, select some property of a current-carrying wire and then decide what to call the current when this property has such and such a value. We might decide that if the paddle-wheel you saw in the TV demonstration takes a defined number of seconds to move a defined distance, then one coulomb of charge passes through the tube every second. However, it is difficult to obtain reproducible results with such a tube. The internationally agreed definition of the unit of current in fact makes use of the forces between current-carrying wires. In effect, it prescribes what we are to take as the magnitude of the currents once we have measured the force between two short lengths of wire at a fixed position relative to each other. Force measurements are easily made with a newton balance (which has, in its calibration, involved the basic units of mass, length and time). Furthermore, such balances give closely reproducible results. Here is an example of how the international definition works in practice with one particular experimental layout.

We wish to specify the current, i, flowing in a wire in terms of the force between this wire and an adjacent current-carrying wire. Perhaps the simplest arrangement is to bend a single wire back on itself so that there are two parallel wires, each carrying the current i. This layout is shown in Figure 3 (p. 16), where the wires run on sliders to prevent the bending of the wires which occurred with the circuit of Figure 2. Newton balances connected to the wires enable the force of attraction between them to be measured accurately. *With this particular geometrical arrangement* it has been agreed that if there is a force of 2×10^{-7} N per metre of length between two infinitely long straight parallel conductors of negligible cross-section placed 1 m apart in vacuum, then there is a *defined* current of 1 coulomb of charge (written 1 C) passing a point in each wire per second. If the force is 8×10^{-7} N per metre the current is 2 C per second.

If it is 18×10^{-7} N per metre the current is 3 C per second. As these examples show, the force is taken to be proportional to the product of the currents in each wire; in this particular layout the currents will necessarily be equal.

Because of the conventions of the subject, the directions indicated by the arrows are in fact in the opposite sense to the direction of movement of the paddle-wheel seen in the TV demonstration. Before experiments like the paddle-wheel were performed, one could only guess the direction of the current. In fact, the wrong sense was originally chosen, but this wrong choice is still retained in indicating the current direction.

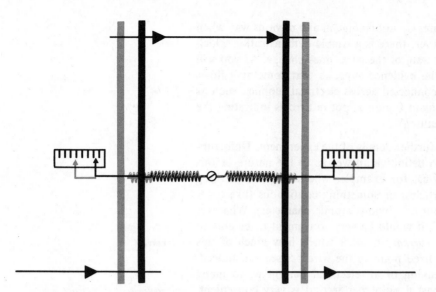

Figure 3 *A schematic current-balance. The force of attraction between the two wires, each carrying the same current,* i, *is measured by a newton balance.*

You should not attempt to memorize the particular magnitude of the force that must exist between the two wires for the current to have a defined value of 1 coulomb per second, any more than you should try and memorize the number of wavelengths involved in the definition of the metre. You should, however, appreciate that the definition of the coulomb is every bit as fundamental as those of the metre, the kilogramme, and the second. All physically measurable quantities can be expressed ultimately in terms of suitable multiples of the basic units of mass, length, time and charge.

Equal currents flow in each of two long parallel conductors placed 1 m apart in vacuum. If the force between the wires is 160×10^{-7} N when measured over a 5 m length what current flows in each wire?

The force per unit length of wire is $160 \times 10^{-7}/5 = 32 \times 10^{-7}$ N, which is 16 times the force produced by unit current. Therefore, as the force is proportional to the square of the current in one of the wires (which is the same thing as the product of the currents in each wire), $\sqrt{16} \times 1$ C $= 4$ C flows past a point in each wire per second. As it is rather tedious to have to keep repeating 'a coulomb of charge passing a point in the wire per second', this is conventionally shortened to *ampere*, or *amp* (written A). So in this example the current is 4 A. While it is possible to measure such small forces as 10^{-7} N, sensitive balances are required. Practical current meters measure not the force between two single lengths of parallel wires but the force between two coils of wire (each consisting of many such parallel wires all carrying the same current). Figure 4 shows such a meter or 'current balance', where the interacting coils are placed one inside the other (one of the coils is removed in Figure 4). As the coils are connected up one after the other, in *series*, the same current flows through each one. The direction of the current is such that the pair of coils at one end of the balance attract, while the pair at the other end repel, with the result that the beam carrying the coils rotates. To bring the beam back to the horizontal, weights are hung at appropriate places on the beam. From the position of these weights, the force acting between the coils can be calculated and hence the current can be deduced.*

ampere

* The international definition relating current to force does not attempt to cover every possible type of wire layout. Instead, it deals with two short lengths of current-carrying wire. To discuss any practical layout one must 'integrate' the relation dealing with the short lengths of wire. We have quoted the integrated relation in the case of the two long parallel wires.

16

Figure 4 *A practical current-balance. A coil is hung from each arm of the balance and the forces between these coils and fixed outer coils (one of which is removed) are attractive at one end of the balance and repulsive at the other. Weights are added to the beam of the balance so as to return it to the horizontal position.*

current balance

The type of 'ammeter' most commonly found in science laboratories measures not the force between two current-carrying coils but the force between one such coil and a magnet. As will be shown in Unit 23 a magnet, in fact, behaves like a current-carrying coil. This type of meter must, however, be calibrated against a current balance. Figure 5 shows a close-up of such a meter.

ammeter

In all these experiments involving current-carrying wires, the electric charge has been moving relative to the person carrying out the experiment or, put more formally, relative to the force-measuring device. Such forces may, for obvious reasons, be labelled as *electrodynamic*, although, for historical reasons, they are more usually called *magnetic* forces. However, when charges are at rest relative to the experimenter, an apparently different force appears; the so-called *electrostatic* force. The interrelation of these two forces is discussed further in the radio programme of this Unit.

electrodynamic force

electrostatic force

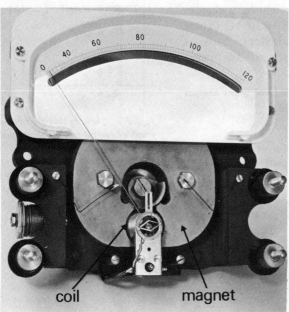

coil   magnet

Figure 5 *A practical ammeter. The force between the current-carrying coil and the magnet causes the coil to rotate about the axis. A pointer is attached to the coil.*

17

In the television programme you saw a demonstration of electrostatic forces. Using the circuit of Figure 6, the experiment showed that charge could be collected on metal spheres, or rather that charge could be transferred from one sphere to another. Remember how, when the switch S was closed, a current did appear briefly in the circuit and as a result of this transfer of charge, the upper sphere (suspended from a spring) was attracted towards the lower one. Moreover, *the spheres continued to be attracted even when the current had ceased flowing* and the switch had been reopened, so the attractive force cannot be electro-dynamic in origin. Nor can it be gravitational in origin as the masses of the spheres have not changed perceptibly.

We have observed a new force, the electrostatic force. What can be found out about it? How, for instance, does the force between two spheres depend on the magnitude of the charge on each sphere and their separa-tion? To make the investigation, we will have to use a set of spheres with different charges. These can be simply obtained by repeating the experi-ment you saw on TV (Fig. 6), but with a different number of batteries in the circuit. With more batteries in the circuit the area under the graph of current against time is greater. In any particular experiment the charge which is transferred from one sphere to another is represented by *the area under the graph* of current against time. To see why this is so we approxi-mate the smooth plot of current against time (as plotted out automatically on the cathode-ray oscilliscope) by a succession of steps (Fig. 7). We suppose that during an interval δt the current is constant with a value I, that is, there is I coulomb of charge per second flowing from one sphere to the other during this interval. If we then multiply the current I (in C s^{-1}) by the time, δt (in s) for which it flows, we get the total charge δQ (in C) flowing in this interval. (E.g. if $\delta t = 10^{-6}$ s and $I = 5 \times 10^{-3}$ C s^{-1}, then $\delta Q = I \times \delta t = 5 \times 10^{-3}$ C s$^{-1} \times 10^{-6}$ s $= 5 \times 10^{-9}$ C.)

But the product of I and δt is nothing more than the area under this particular step (the height times the base). The same argument holds for each and every step, so the total charge transferred in the experiment from one sphere to the other is the total area under the curve, which can be measured (e.g. by the 'counting squares' technique).

By repeating the charging experiment with different numbers of batteries in the circuit, we can acquire a large selection of spheres of known charge. In all experiments of this kind, one of the spheres is always found to have lost charge, the other to have gained it. Those spheres which have gained charge are conventionally labelled positive (+), those which have lost charge are labelled as negative (−).*

A collection of charged spheres obtained in this way might have charges of say $+1.5 \times 10^{-8}$ C, -1.5×10^{-8} C, $+3.0 \times 10^{-8}$ C, -3.0×10^{-8} C. (These are the sort of charges you collect with a few hundred car batteries connected to spheres with the diameter of about 0.1 m placed 0.3 m apart.)

In Unit 2 it was stated that like charges repel each other and unlike charges attract. How might we verify this?

We could hang a sphere of one sign charge at the end of a newton balance and bring up another of similar or opposite sign and watch the result.

How should we now proceed with a systematic study of how the force F between two spheres depends on their charges, Q_1 and Q_2, and on the separation r between their centres?

* *As already mentioned (p. 15), the actual charge carriers really move in the opposite sense to that suggested by the conventions. These carriers (electrons) possess a charge which is taken to be negative. A gain of positive charge is therefore, strictly speaking, better described as a loss of negative charge.*

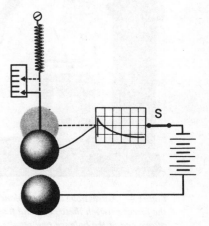

Figure 6   *Electrostatic forces. On closing the switch S the two metallic spheres attract each other. The cathode-ray oscilloscope enables the charge transferred from one sphere to the other to be measured.*

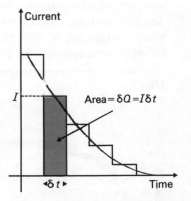

Figure 7   *Showing that the area under a graph of current against time equals the charge δQ transferred. In an interval δt the charge transferred is $I\delta t$ where I is the current; $I\delta t$ is the magnitude of the shaded area.*

If you recall the techniques used in studying gravitational forces (section 4.2.1) you may have the answer. In exact analogy with our studies of gravitational forces, one finds out how F depends in turn on Q_2 (while the values of Q_1 and r are kept constant) and finally on r (while the values of Q_1 and Q_2 are kept fixed). The combined results of these experiments show that . . .

$$F \propto Q_1 \, Q_2/r^2$$

or
$$F = \left(\frac{1}{4\pi\epsilon_0}\right) Q_1 \, Q_2/r^2 \ldots\ldots\ldots(5)$$

where the constant of proportionality has been written as $1/4\pi\epsilon_0$ (ϵ is pronounced 'epsilon'). The constant is given this rather awkward form because it simplifies the notation in more advanced problems. So long as ϵ_0 is a constant, $1/4\pi\epsilon_0$ will, of course, be constant.

Since the experiments have employed known charges at known separations and the resulting forces have been measured, this data is sufficient to determine $1/4\pi\epsilon_0$ just as the gravitational data determined G.

> With two spheres of charge $Q_1 = Q_2 = 1.4 \times 10^{-8}$ C (deduced from the area under the graph of current against time) $F = 0.217 \times 10^{-5}$ N when $r = 0.90$ m. Deduce $1/4\pi\epsilon_0$.

Unlike G, the value of which is apparently uninfluenced by the medium in which the experiment is carried out, the electrostatic constant of proportionality varies from medium to medium. For example, if the surrounding of two charged but otherwise isolated spheres is changed from air to glycerine, the force between them is reduced by a factor of 42.5.*

From equation 5

$$\frac{1}{4\pi\epsilon_0} = \frac{Fr^2}{Q_1 \, Q_2}$$

$$= \frac{0.217 \times 10^{-5} \times (0.9)^2}{1.4 \times 10^{-8} \times 1.4 \times 10^{-8}} \frac{\text{N m}^2}{\text{C C}}$$

$$\therefore \frac{1}{4\pi\epsilon_0} = 8.99 \times 10^9 \text{ N m}^2 \text{ C}^{-2}$$

Since N means kg m s^{-2} we can, if we wish, write

$$\frac{1}{4\pi\epsilon_0} = 8.99 \times 10^9 \text{ kg m}^3 \text{ s}^{-2} \text{ C}^{-2}$$

4.2.3 Nuclear forces

In later Units we shall be presenting evidence that all matter is made up of atoms, and that these atoms have their own complex structure, consisting of a central *nucleus*, with a diameter of some 10^{-15} m, surrounded by an electron cloud which extends out to a diameter of around 10^{-10} m. We shall be presenting evidence that the nucleus consists of roughly equal numbers of protons and neutrons. Protons are particles with a charge of the same magnitude as that of the electron but of opposite sign and with a mass of 1.67×10^{-27} kg, which is nearly 2 000 times the mass of the electron. Neutrons are uncharged particles with approximately the same mass as protons. How, we may ask, can such a nucleus remain stuck so firmly together? Surely the protons will be driven apart out of the nucleus by electrostatic repulsion? Not necessarily. Perhaps the force of gravitational attraction, F_g say, between the two protons is at least as great as the force of electrostatic repulsion, F say. We can check this by seeing if F_g/F_e has a value of unity or greater.

atom

nucleus

proton

neutron

---

* In terms of equation 5 this means that the constant of proportionality is reduced by a factor of 42.5. To cover such general cases equation 5 is normally written as

$$F = \left(\frac{1}{4\pi\epsilon\epsilon_0}\right) Q_1 \, Q_2/r^2$$

The constant ϵ is known as the dielectric constant *or* relative permittivity *of the medium. The value of ϵ for glycerine is therefore 42.5. Of course, in a vacuum ϵ has a value of 1 exactly. In air, $\epsilon = 1.0005$.*

relative permittivity

Treating the protons as spherical particles of mass m and charge e, write down expressions for F_g and F_e. Next evaluate the ratio F_g/F_e.

From equations 4 and 5:

$$F_g = Gm^2/d^2, \text{ where } d \text{ is the separation between two protons.}$$

and

$$F_e = \left(\frac{1}{4\pi\epsilon_0}\right) e^2/d^2$$

So

$$\frac{F_g}{F_e} = \frac{Gm^2}{d^2} \Big/ \frac{e^2}{4\pi\epsilon_0 d^2}$$

$$= \frac{Gm^2}{d^2} \times \frac{4\pi\epsilon_0 d^2}{e^2}$$

$$= 4\pi\epsilon_0 Gm^2/e^2 \dots\dots\dots\dots\dots\dots\dots\dots\dots\dots(6)$$

Notice how the d^2 cancels out; a direct consequence of the fact that the gravitational and electrostatic forces have the same spatial dependence.

So this argument does not depend on the separation between nuclear protons.

Substituting into equation 6 the values:

$$\frac{1}{4\pi\epsilon_0} = 8.99 \times 10^9 \text{ N m}^2 \text{ C}^{-2}$$

$$G = 6.67 \times 10^{-11} \text{ N m}^2 \text{ kg}^2,$$

$$e = 1.60 \times 10^{-19} \text{ C},$$

$$m = 1.67 \times 10^{-27} \text{ kg},$$

gives:

$$\frac{F_g}{F_e} = \frac{6.67 \times (1.67 \times 10^{-27})^2}{8.99 \times 10^9 \times 10^{11} \times (1.60 \times 10^{-19})^2} \frac{\text{N m}^2 \text{ kg}^{-2} \text{ kg}^2}{\text{N m}^2 \text{ C}^{-2} \text{ C}^2}$$

i.e.

$$\frac{F_g}{F_e} \approx \frac{1}{10^{36}}$$

The repulsive force is some million million million million million million times larger than the attractive force. Without any doubt some new basic force operates inside the nucleus! Clearly it must be an attractive force. It must also be short-ranged, i.e. it must not extend far outside the nucleus, for if it were to extend further we would be unaware of the much weaker gravitational and electromagnetic forces in the world around us. (In fact, within the short range of the nucleus, this force is some hundred times stronger than the electromagnetic force.) Atomic nuclei have neutrons as well as protons in them and are *stable*; that is, they do not in general break up spontaneously, so the nuclear force must act on the neutrons as well as the protons. Whether the nuclear force acts just as strongly on neutrons as it does on protons remains to be seen (in Unit 31).

The force, some of whose properties we have just listed, is usually referred to as the nuclear *strong interaction*. To account for certain other aspects of the behaviour of sub-nuclear *elementary particles* such as, for example, the process leading to the decay of the muon, it is necessary to postulate yet another basic force, the *weak interaction*. Discussion of this is deferred until Unit 32.

strong interaction

weak interaction

4.3 Fields

4.3.1 The field concept

You will recall how, in Unit 2, we introduced the concept of a *field* as a means of accounting for action at a distance. An analogy was drawn with the behaviour of a stretched rubber membrane; distort it in one place and it distorts all around. But a field, remember, is only a model, introduced to explain why, for example, one charge should attract or repel another one placed some distance away.

Here is yet another model of a field which presents an alternative viewpoint and serves to recall the salient points of the earlier discussion in Unit 2.

If you were to take two corks and place them on a pool of water and were then to wiggle one cork up and down, it would not be long before the second cork also wiggled up and down.

Why?

It is a trivial question, because in our imaginations we can see that what happens is that one cork disturbs the water and the second one experiences the disturbance. But now suppose that bad eyesight prevents you seeing the water.

How would you describe the experiment?

You would see one cork being moved up and down and a second one going up and down in sympathy. You would conclude that there is a direct interaction between the corks; action at a distance.

Hold this text at arm's length. Certainly you can feel the Earth pulling on it. Perhaps you conclude that there is a direct interaction between the Earth and the book. In imagination, watch a couple of current-carrying wires in close proximity to each other—a direct interaction? Those charged spheres—are they really acting on each other at a distance? Might it not simply be that we are blind to these forces? Perhaps what really happens is that the Earth creates some disturbance which then interacts with the book. Perhaps one current-carrying wire creates a disturbance which the other current-carrying wire senses. Perhaps one charged sphere creates a disturbance which the second one merely responds to. The agony of these speculations is that, like A. A. Milne's Pooh, we never can tell. If you have poor eyesight, the only way to find out if the water is disturbed is to place something on it and look for an interaction. The only way we can detect whether a mass, a current-carrying wire, or a charged sphere produces any disturbance is to introduce another mass, another current-carrying wire, or another charged sphere. But we could just be witnessing a direct interaction—we never can tell.

No one is going to dictate your choice of a field model. In the past, such models have included springs, bent tubes, and even gear wheels that fill space. Of course, the more you use any one model, the more deeply convinced you may become of its reality. Keep several at your finger tips and you may avoid falling into this trap. The following experiment was originally designed to demonstrate to schoolchildren just how a particular model, that of a stretched spring, can account for the forces of gravitational attraction. It can be such a convincing demonstration that having performed the experiment one can readily refute all suggestions that gravitational attraction just might not, after all, be due to a stretched spring. As a tale with a moral, the experiment is well worth performing, in spite of it being of the nature of a children's game.

> **First decide to accept the model that the interaction between the Earth and say a book is attributable to an invisible spring. Find a massive book and place it on the palm of your hand, keeping your arm fully outstretched. Now close your eyes and lift the book at arm's length, saying as you do so, 'I am stretching a spring'. As you lower the book say, 'I am relaxing the spring'. Keep this up for some time and you are likely to become deeply convinced of the reality of the spring connecting the book to the Earth.**

It is only an experiment in self-hypnosis, but despite this, or rather because of this, it makes a worthwhile comment on the reality of models.

4.3.2 The electric field

If we are to make full use of the concept of a field, we must put it on a more quantitative basis. Fortunately, while there are diverse qualitative pictures of what a field is 'like', there is a fair amount of agreement among scientists as to how to express field magnitudes (strengths).

It is an observed fact that a sphere possessing a charge Q will exert a force of magnitude of $Qq/4\pi\epsilon_0 r^2$ on a sphere of charge q placed a distance r away. It is known too that the force is directed along a line connecting the centres of the spheres, that it is attractive when Q and q are of opposite sign and repulsive when Q and q have the same sign. Let us imagine that what actually happens is that one charge creates a field with which the second one subsequently interacts. Figure 8 shows a schematic diagram of the field. If you like, you may think of the lines as springs waiting to push on any other charged body. You may even care to imagine that introducing this other charged body (which of course will have its own spring system) cuts the tapes which prevent the springs acting on nothing. The mechanism is your choice.

Mechanisms apart, we must agree on a definition for the value of the field at, say, a distance r from a charge Q. Following tradition, the electric field E at any point is defined as the *would-be force per unit positive charge placed at the point*. We have taken the somewhat unusual step of incorporating the words 'would-be' to emphasize that we are assuming the disturbance is still there even when we are not sampling it. 'Per unit positive charge' means that we calculate what the force would have been if we had placed a charge of $+1$ C at the point in question. For example, if we found a force of 10^{-2} N on a charge of $+10^{-3}$ C the force on $+1$ C would have been 10^{-2} N$/10^{-3}$ C $=10$ N C^{-1}. It is like calculating the speed in metres per second of a car that has been observed to travel 3 m in 10^{-1} s; its speed is 3 m$/10^{-1}$ s $=30$ m s^{-1}. In representing the field schematically, as in Figure 8, it is usual to draw in arrow heads to show the field direction, i.e. the direction of the force on the positive 'test charge' introduced at the point.

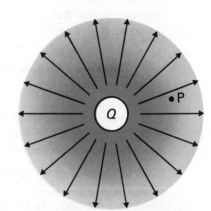

Figure 8 *A representation of an electric field produced by a charge* Q. *The lines represent the field. The arrow indicates the direction of the field.*

From our definition, it follows that the electric field, E, at a point P, at a distance r from a single charge Q, is

electric field

$$E = \frac{F}{q}$$

22

where F is the force that a charge q would experience when introduced at point P. Substitution for F from equation 5 gives

$$E = \frac{Qq/4\pi\epsilon_0 r^2}{q}$$

i.e.
$$E = \frac{Q}{4\pi\epsilon_0 r^2} \dots\dots\dots\dots\dots(7)$$

What are the units of E?

From the definition, it should be 'newton per coulomb', i.e. the force per unit charge. Remembering that $1/4\pi\epsilon_0$ has units of N m² C⁻² (see p. 19) the right-hand side of equation 7 has units of (N m² C⁻²) . (C) . (m⁻²) which are indeed N C⁻¹.*

SAQ 6
Calculate the electric field at a distance of 2×10^{-10} m from a charge of 1.6×10^{-19} C.

The problem is worked out on p. 46.

You should obtain an answer of 3.6×10^{10} N C⁻¹. Perhaps you recognized the figures in the problem. What you have just calculated is the field produced by a single electronic charge at a distance away of about one atomic diameter.

SAQ 7
What will be the force on a 1.6×10^{-19} C charge when introduced into a region where the electric field is 3.6×10^{10} N C⁻¹?

The problem is worked out on p. 46–47.

It is around 5.76×10^{-9} N.

This is actually the sort of attractive force that one might expect to find between two atoms in a molecule of common salt (sodium chloride). As you will learn in Unit 9, in sodium chloride the sodium atom has transferred an electron to the chlorine atom (which therefore has a charge of magnitude 1.6×10^{-19} C.) Using relatively simple concepts, we have made a quite fundamental, if rough, calculation of the attractive forces to be found in one particular, but important, molecule.

What charge configuration produces a field of 10³ N C⁻¹?

There is no unique answer. Remembering the form of equation 7 and that $1/4\pi\epsilon_0 = 9 \times 10^9$ N m² C⁻² (near enough) this field could be produced for example by a single charge of $+1$ C at a distance of 3×10^3 m away. Alternatively the field could be produced by four times the charge twice as far away. Or by several charges. You never can tell. Field values are experimentally measurable quantities; we do not have to know what charge configuration is producing these fields. Conversely we cannot infer the configuration uniquely.

---

* *Later in the Unit, we shall be introducing the term volt (V) as shorthand for N m C⁻¹. Therefore the units of field, namely N C⁻¹, can also be written as V m⁻¹.*

4.3.3 The gravitational field

In exact analogy with the electrostatic case, one can define the gravitational field E_g at a point as the *would-be force per kilogramme mass placed at a point*.

From this definition write down an expression for the gravitational field E_g at a distance r from a spherical mass M.

The answer, obtained by writing down the force which would act on, say, a mass, m, at the point in question (using equation 4) and then dividing the result by m to give the force per unit mass:

$$E_g = GM/r^2 \dots\dots\dots\dots\dots(8)$$

gravitational field

Given that the Earth has a mass of about 6×10^{24} kg and a radius of about 6.4×10^6 m, deduce the gravitational field at the Earth's surface.

This field is 9.8 N kg^{-1}. This value, calculated by substituting $G = 6.67 \times 10^{-11}$ N m^2 kg^{-2}, $M = 6 \times 10^{24}$ kg, and $r = 6.4 \times 10^6$ m into equation 8, differs slightly from the field that would be measured by, say, hanging up a 1 kg mass at the end of a newton balance. The discrepancy between the calculated and observed field values is attributable in part to the centrifugal force acting on the test mass as a result of the rotation of the Earth. This discrepancy will be discussed further in Unit 22.

4.3.4 The magnetic field

magnetic field

In giving field representation to the interaction between two current-carrying wires, it is usual to say that one of the wires produces a magnetic field which can be represented by a series of concentric rings as shown in Figure 9. The arrows, indicating the direction of the field, are drawn in such a sense that a right-hand screw (e.g. a corkscrew) rotated in this sense would advance in the direction of the current flow. The reasons for drawing rings rather than, say, radial lines are partly historical.*

The magnitude of the magnetic field at some point, P, in Figure 9 is defined as the 'would-be force per metre of wire' carrying a current 1 A when placed parallel to wire W. In a more general situation, the field magnitude is found from the maximum force on the current-carrying element (the test wire if you like) as its orientation is altered. We will have more to say about magnetic fields in later Units. Indeed, measurement of the Earth's magnetic field provides useful clues as to the Earth's structure.

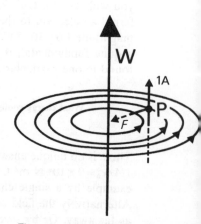

Figure 9   A representation of a magnetic field produced by a current-carrying wire, W, showing the magnetic force F acting on a short length of wire carrying a current of 1 ampere. Note that the force is at right angles to both the current and the magnetic field, i.e. it is directed radially inwards.

* If a magnetic compass is moved in a circular path around such a wire, the axis of the magnet keeps tangent to the circle. This does not however prove the existence of such a field, for the magnet itself contains 'circulating' charges at the atomic level. The behaviour of the compass can be explained equally well in terms of a direct 'action at a distance' between the current in the wire and the 'currents' in the magnet.

4.4 Energy

4.4.1 A partial picture

In all the experiments with forces that we have described, we have taken it as axiomatic that such forces are, so to speak, available gratis. But we know as a matter of experience that we cannot, for example, keep on digging in the garden, i.e. keep on applying a force causing things to move, unless we consume fuel in the form of food. Nor is the requirement peculiar to humanity. Use a rotary cultivator if you like, but it requires petrol. An electric cultivator requires connecting to the mains. Operate such an electrical device and more fuel, perhaps coal, will be consumed in the power station. The performance of any useful everyday job of work in which a force is made to move an object (or, put more formally, in which a force moves its point of application) requires fuel in one form or another. This something, which is stored in fuels and which enables useful jobs to be done, is called *energy*. In the present context, 'useful jobs' are defined as those in which the force moves its point of application. The word fuel must not be thought of too restrictively. It might, for example, be possible to dig the garden by means of a Heath Robinson device in which a weight hung over a nearby cliff descends towards sea-level. In this case the 'fuel' is to be sought in the Earth's gravitational field; if you adopt a 'spring' model for gravitational fields the energy is stored in the stretched spring. But what happens to the energy which is contained in fuels after the fuels have been used and the job has been done? One line of investigation is to look at the 'end products'; if they can themselves do useful jobs then, by definition, at least some of the energy present in the fuel has not vanished into nothingness. To throw a puck, for example, one requires chemical energy in the form of food. Now consider the moving puck—it could be made to collide with a spring. The spring would be compressed and a compressed spring can do many useful jobs. So the moving puck had energy, *kinetic energy* as we say, energy by virtue of its movement, and this was transformed into *potential energy* in the spring, energy by virtue of the spring's configuration. But, instead of compressing a spring, the puck could have hammered a nail into a piece of wood. Some of the puck's kinetic energy would have been transformed into potential energy in the strained wood but some of it would have appeared as heat or *thermal energy*. The nail feels warmer than its surroundings and, in principle, there is no reason why this heat should not be utilized. Indeed, if you follow through any sequence of energy transfers —such as a compressed spring which works a dynamo which drives a light bulb, for example—you will always find that at each stage part of the energy is transformed into thermal form. Once the thermal energy has 'run away' into the surrounding cooler region it is no longer accessible unless we have something cooler still to transfer it to. Steam can only be used to drive a steam engine, for example, if there is a cool region for the steam to condense in. The tepid water which results might still be used to do other useful jobs, but cool water at the same temperature as the air in the room can by itself do no further useful jobs. This is not to say that the energy has disappeared; it is just inaccessible. Indeed, one naturally assumes that energy must be conserved: it must 'go somewhere'. If it does not turn up in an obvious form, one starts offering excuses in the form of new names for the 'missing' energy.

energy

kinetic and potential energy

thermal energy

conservation of energy

After a rotary cultivator has dug a garden, what has happened to the energy that was in the petrol which has been consumed?

A good deal, perhaps about half, of the energy has gone into thermal form. But the general level of a dug garden is higher than an undug one; there is gravitational potential energy in the raised earth—one could arrange for further useful jobs to be done as the earth sinks down to its former level.

4.4.2 Measuring energy transfers

Is there any way of combining the forces which appear when mechanical jobs of work are done with the distance through which these forces act, to obtain a realistic measure of the energy transfer. Should we perhaps multiply the square of the force by the cube of the distance moved, or should we . . . ?

The best way to find out is by experiment.

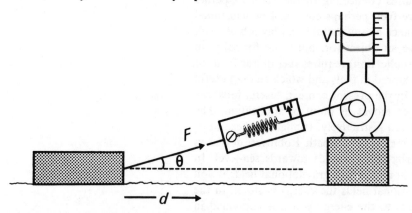

Figure 10 *Measuring energy transfers. A petrol engine fitted with a graduated tank pulls a load across rough ground. The experiment studies how the petrol consumption V depends on F, d and θ.*

Have a look at the apparatus shown in Figure 10. Here a petrol engine fitted with a graduated tank slowly pulls a load across rough ground, via a newton balance, transferring energy in the process *from* chemical form in the fuel *to* thermal form in the ground. We wish to know how the petrol consumption, or rather the extra volume V consumed over that used when running light, depends on the pulling force F, on the distance d through which it acts, and on the angle θ between the direction of the force and that of the movement along the surface. As we cannot hope to study all this in a single investigation, we first study how V depends on d, keeping F and θ constant. The result is that if d is doubled the petrol consumption is doubled, if d is trebled V is trebled. In general

$$V \propto d, \text{ keeping } F \text{ and } \theta \text{ constant} \dots\dots\dots(9)$$

—a result which should be familiar to any car owner.

Next we arrange that it requires, say, twice the pull to move the load, perhaps by roughening the ground, and measure the petrol consumption. Keeping the same d and θ as before, we change the required pulling force and repeat the experiment with these different values of F. The result:

$$V \propto F, \text{ keeping } d \text{ and } \theta \text{ constant} \dots\dots\dots(10)$$

—again a result which should hardly surprise us.

Lastly, we keep F and d constant and vary θ. We discover that

$$V \propto \cos\theta, \text{ keeping } F \text{ and } d \text{ constant*} \dots\dots(11)$$

It is certainly true that in the case when $\theta = 90°$, i.e. $\cos\theta = 0$, the load will not move along the surface and so no fuel can be used in doing this particular job.

Combine the result of the three separate experiments, equations 9, 10 and 11 into a single relation.

* *If you have forgotten what cos θ means, refer to MAFS, section 4.A.1.*

Combining equations 9, 10, and 11 gives:

$$V \propto Fd \cos \theta \ldots \ldots \ldots \ldots \ldots (12)$$

the required expression relating the volume of petrol used with F, d, and θ. But the energy content of petrol must be proportional to the volume present. If it were not, we should find, for example, that a car would travel different distances on the second, third, fourth, etc., gallons of a tankful. We may therefore write:

The energy transferred from the petrol $\propto$ Volume V of petrol used.

Or, from equation 12:

The energy transferred from the petrol $\propto Fd \cos \theta$.

To complete the story, we must discover how the energy gained by the ground is related to the energy lost from the petrol. The two are actually proportional. When the petrol consumption is doubled, twice the number of useful jobs can be done on the now warm surface across which the load has been pulled. Useful jobs in this context might be boiling beakers of water, melting squares of jelly, even frying eggs on the surface. We therefore conclude that:

$$\text{Energy transfer } \textit{from} \text{ the petrol } \textit{to} \text{ the ground} \propto Fd \cos \theta \ldots (13)$$

By convention the constant of proportionality is taken as unity, so

$$\text{Energy transfer} = Fd \cos \theta \ldots \ldots \ldots \ldots \ldots (14)$$

In the SI system, the energy transfer will be measured in units of N m. However, for convenience, N m is shortened to joule (J). Equation 14 may be written somewhat differently by noting that $F \cos \theta$ is the component F_d of force F in the direction of movement of the load. (See section 4.D.7 of *MAFS*.)

joule

$$\text{i.e. Energy transfer} = F_d d \ldots \ldots \ldots \ldots \ldots (15)$$

In the past it used to be fashionable to talk about 'doing work' when referring to energy transfers; the left-hand side of equation 15 would have been called the 'work done'. While there is nothing wrong with speaking of 'doing work', and we use the words ourselves, there are situations where 'doing work', with all its human connotations, just does not ring true. It is easier on the tongue to say, for example, that energy is transferred from chemical form in a battery to thermal form in a wire connected across the terminals than to say that the battery *does work* on the wire.

SAQ 8

Make an order of magnitude calculation of the (chemical) energy required to push a car by hand along a level road through a distance of 1 mile. Assume the car is put in 'neutral'. (As a clue, first try and decide, to the nearest power of ten, what force is required to keep a car moving.) Ignore the extra effort required to get the car going.

The problem is worked out on p. 47.

The answer is about 10^5 N m, i.e. 10^5 J.* This will be a minimum estimate of the energy required, for our bodies are nothing like a hundred per cent efficient at converting chemical energy to mechanical energy.

The next example not only provides further practice in calculating energy transfers; it demonstrates how the energy transferred when a force moves its point of application can be calculated graphically. This graphical technique will be employed in subsequent sections.

* In the past 4.18 J used to be called 1 calorie so the answer might be written as $10^5/4.18 = 2.4 \times 10^4$ calorie [the Calorie (notice the capital letter) referred to in dietetics is 1 000 calories].

calorie
Calorie

A 'black box' has a piece of string coming out of it. You pull on the string via a newton balance, and find that your pull varies with the length of the string pulled in the way shown in Figure 11. How much energy has been transferred from you to whatever is inside the box during the experiment?

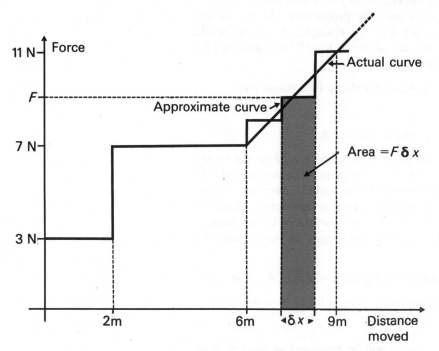

Figure 11 An example of an energy transfer. When a piece of string coming out of a black box was pulled, the pulling force varied with distance pulled in the manner shown.

Between 0 and 2 m the energy transferred is, from equation 15, equal to 3 N × 2 m = 6 N m = 6 J, which you should note is also equal to the area under this portion of the force-distance curve. From 2 to 6 m the energy transferred is 7 N × 4 m = 28 J; again the area under this section of the force-distance curve. To evaluate the energy transferred in the section from 6 m up to 9 m, we approximate the changing force as closely as we wish by a succession of steps. The energy transferred in pulling the string through a distance δx is $F\delta x$, where F is the force required to pull the string. This product $F\delta x$ is the area under this step, and is shown shaded in Figure 11. Repeat the argument under each step, add up all the areas and you will have proved that the total energy transferred between 6 and 9 m is the area under this section of the plot. Indeed the total energy transferred throughout the experiment is the total area under the force-distance curve. The area under the curve between 6 and 9 m is equal to the area of the rectangle of height 7 N and of length 3 m, plus the area of the triangle, of height 4 N and of base length 3 m;* i.e. the energy transferred between 6 and 9 m is 21 J + 6 J. Hence the total energy transferred from the puller into the box is 6 J + 28 J + 27 J = 61 J.

4.4.3 Potential energy

Some types of energy transfer occur so frequently that it is worthwhile making once-for-all calculations. A particularly important example is the energy required to bring one body up from infinity to a point at a given distance from another body which attracts or repels it in a well-defined manner. When the body arrives at this point, it is said to have potential energy (P.E.) which is the energy required to bring the body up from infinity to the point. The potential energy is labelled as positive if the 'pusher' must provide energy in bringing the body up, and negative if energy is transferred to the 'pusher' in the process. The concept of potential energy is not a simple one, so you should not be unduly worried if it eludes you on a first reading.

potential energy

* *The area of a triangle is ½ (base × height). See MAFS, section 2.A.2.*

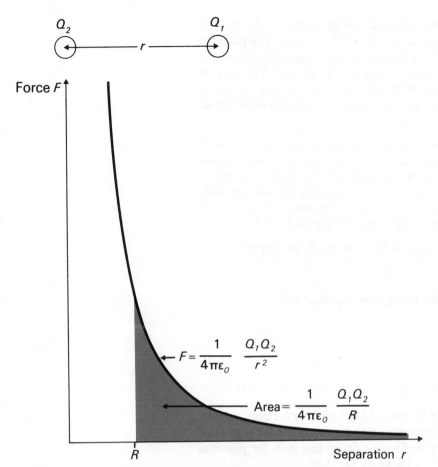

Figure 12 *Electrostatic potential energy. The shaded area represents the energy required to bring Q_1 up from infinity to within a distance R of Q_2, i.e. it represents the potential energy of Q_1.*

The two spheres of Figure 12 carry charges Q_1 and Q_2 respectively. As Q_1 approaches Q_2 the magnitude of the force F which Q_1 exerts on Q_2, when they are separated by a distance r, is given by equation 5 namely $F = Q_1 Q_2/4\pi\epsilon_0 r^2$. The variation of F with r is shown plotted in Figure 12. From what we have just learnt in the worked example of section 4.2.2, the total energy required to bring Q_1 from infinity up to a distance R from Q_2 is the area under the force-distance curve between R and infinity, and this is shown shaded in Figure 12. While anyone familiar with elementary integral calculus may be able to work out the result for themselves (a derivation is given in Appendix 1 (Black)), we can at this stage only state that

the area under the curve between R and infinity $= \dfrac{Q_1 Q_2}{4\pi\epsilon_0 R}$ ....(16)

and consequently

the P.E. of Q_1 when a distance R from Q_2 $= \dfrac{Q_1 Q_2}{4\pi\epsilon_0 R}$ ....(17)

Note, and this is important, that the denominator of equation 17 contains

R, not R^2.

Satisfy yourself that the units of the right-hand side of equation 16 or 17 are those of energy.

In the SI system, Q_1 and Q_2 are in C, $1/4\pi\epsilon_0$ is in N m^2 C^{-2} (see p. 19), and R is in m. Therefore the right-hand side of equation 17 will be in C^2 N m^2 C^{-2} m^{-1}, i.e. N m or J, which is indeed the SI unit of energy.

SAQ 9
What energy is required to bring a charge of -1.6×10^{-19} C up from infinity to a distance of 2.5×10^{-10} m from a charge of $+1.6 \times 10^{-19}$ C?

The answer is worked out on p. 47.

By definition, this energy is the potential energy of the -1.6×10^{-19} C charge when 2.5×10^{-10} m from the $+1.6 \times 10^{-19}$ C charge. The potential energy works out as -9.2×10^{-19} J. The negative sign indicates that energy has been transferred from the electric field to the 'pusher'. You may be interested to know that in this particular example we have, in effect, estimated the energy which is given out when a sodium chloride molecule is formed.

You will frequently hear people refer to the *electrical potential* at a point. By this they mean the *energy required per coulomb of positive charge brought up from infinity to the point*. To calculate the electrical potential $V(R)$ at a point a distance R away from a charge Q_2, we must calculate the energy required to bring up a charge Q_1 to this point and then divide the answer by Q_1 to obtain the energy required per coulomb.*

electrical potential

i.e. $\quad V(R) = \dfrac{\text{Energy required to bring } Q_1 \text{ from infinity to the point}}{Q_1}$

i.e. $\quad V(R) = \dfrac{\text{P.E. at } R}{Q_1}$ or, substituting from equation 17,

$$V(R) = \left. \frac{Q_1 Q_2}{4\pi\epsilon_0 R} \right/ Q_1$$

i.e. $\quad V(R) = \dfrac{Q_2}{4\pi\epsilon_0 R}$ $\quad \dotfill (18)$

Electrical potential is measured in units of J C^{-1} but this is conventionally shortened to *volt* (V). You will remember that J is shorthand for N m while N is shorthand for kg m s^{-2}. Therefore the equivalent units for potential are N m C^{-1} or (kg m s^{-2}) m C^{-1}, i.e. kg m^{-2} s^{-2} C^{-1}.

volt

Another term you will often hear used is the electrical *potential difference* (P.D.) between two points. By this is meant the energy required per coulomb of positive charge moved from one point to the other. If, as in Figure 13, a charge Q_1 is brought up from a distance of, say, R_2 from Q_2 to a distance R_1 from Q_2, the energy required is the shaded area. But this area is only the area under the curve from R_1 to infinity minus the area under the curve from R_2 to infinity. Both of these areas we know from equation 18. Therefore,

potential difference

energy required in bringing Q_1
from R_2 to R_1 $\quad = \dfrac{Q_1 Q_2}{4\pi\epsilon_0 R_1} - \dfrac{Q_1 Q_2}{4\pi\epsilon_0 R_2}$

Dividing through by Q_1 gives:

energy required per unit positive
charge brought from R_2 to R_1 $\quad = \dfrac{Q_2}{4\pi\epsilon_0 R_1} - \dfrac{Q_2}{4\pi\epsilon_0 R_2}$

The left-hand side is, by definition, the potential difference between the two points,

i.e. potential difference between
the points $\quad\quad\quad\quad\quad = V(R_1) - V(R_2) \dots\dots(19)$

The following example shows the usefulness of this apparently abstract calculation.

> A 'battery' has terminals marked O, $+1.5$ V, $+3.0$ V, $+4.5$ V, $+6.0$ V, $+7.5$ V. How much energy is given out if $+1$ C of charge (i.e. 6.7×10^{18} electrons) is allowed to move through a wire connecting the 7.5 V terminal to the 1.5 V one?

---

* Note that V (R) *does* not *mean 'V times* R'. *We call the electrical potential* V (R) *instead of just* V, *to remind ourselves that it depends on* R.

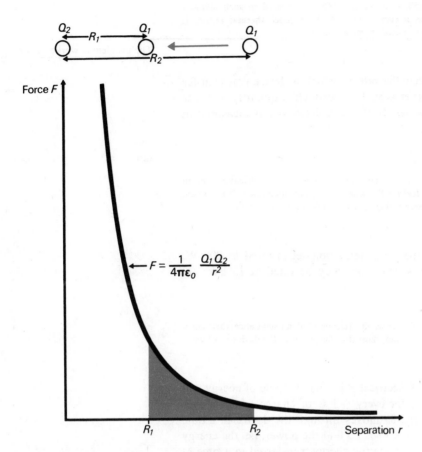

Figure 13 *Illustrating the electrical potential difference between two parts.*

The shaded area represents the energy required to move a charge Q_1 *from a distance* R_2 *to a distance* R_1 *from a charge* Q_2.

The potential difference between these two terminals is, from equation 19, $7.5 \text{ V} - 1.5 \text{ V} = 6 \text{ V} = 6 \text{ J C}^{-1}$. In other words, 6 J of energy is transferred to the wire (as heat) when 1 C of charge flows between these terminals. As the difference between any two adjacent terminals on this battery is 1.5 V we can obtain 1.5 J of heat energy for every coulomb transferred between two adjacent terminals.

> **An electron (charge -1.6×10^{-19} C) is accelerated from rest through a potential difference of 50 V. How much energy has it acquired in the process?**

As the potential difference is the energy transferred per coulomb, the energy acquired by the electron is the potential difference × charge; i.e. $50 \text{ V} \times (1.6 \times 10^{-19})$ C, i.e. $50 \text{ J C}^{-1} \times (1.6 \times 10^{-19}) \text{ C} = 8.0 \times 10^{-18}$ J. If one electron is accelerated through a P.D. of 1 V, it acquires an energy of $1 \times (1.6 \times 10^{-19})$ J. Many scientists use the term *electron volt* as an energy unit, by which they mean the energy acquired by one electron in moving through a potential difference of 1 V, rather as we have used N m or J as our unit of energy. Since 1.6×10^{-19} J is a rather small unit, even for nuclear physicists, they may for example use the term MeV which is the energy required by an electron in moving through a P.D. of 10^6 V. (1 MeV= 1 Mega—electron volt.)

electron volt

MeV

> **What, in joules, is the value of 1 MeV?**

The answer is $10^6 \times (1.6 \times 10^{-19}) \text{ J} = 1.6 \times 10^{-13}$ J. You are not expected to remember these numbers, for you can always look them up in section 3 of *HED*.

31

SAQ 10

SAQ 10
The potential difference between the 'live' and the 'neutral' terminals on a mains plug is typically 250 V in Britain. When connected to such mains a one-bar electric fire takes a current of 4 A. How much thermal energy is transferred to the room per second by such a fire?

The problem is worked out on p. 47.

The answer is 10^3 J s^{-1}. Indeed the rate at which a device can transfer energy from one form to another is such a worthwhile quantity to know that it is given the name of *power*. In the SI units power is measured in J s^{-1}, abbreviated to *watt* (W).

power
watt

Have a look at the manufacturer's label on some pieces of electrical equipment about your home to get a feel for the power of various appliances. It is modern practice to include the power rating on all electrical appliance labels.

You may find, for example, that your television set is rated at 200 W. Your cooker with everything switched on, may be rated at 13 kW, i.e. 13 000 W.

If there is a P.D. of V volt across the terminals of an appliance carrying a steady or 'direct' current of I amp, show that the power of the device is VI watt.

It follows from the definition of electrical P.D. that V Joule of energy will be transferred to the appliance for every $+1$ C of charge which moves through it. Therefore if a charge δq moves through the device in a time interval δt the energy transferred is $V\delta q$. To find the power, i.e. the energy transferred per second, we must divide the energy transferred in a time δt by the time interval δt:

$$\text{Power} = \text{rate of transfer of energy}$$

$$= V\delta q/\delta t$$

Since a charge δq moves through the device in a time δt, $\delta q/\delta t$ is the charge flowing through the appliance per second, i.e. it is the current I.

$$\therefore \quad \text{Power} = VI \dots\dots\dots\dots\dots(20)$$

Equation 20 can only be applied to 'direct current' devices. It cannot, in general, be applied to calculate the power of an 'alternating current' device.*

SAQ 11
What current passes through a 150 W light bulb operating from a 250 V direct current supply?

The problem is worked out on p. 47.

The answer is 0.6 A.

4.4.4 Kinetic energy

Another calculation which is worth making once and for all is the energy transferred to an object as it is accelerated from rest. The puck of mass m in Figure 14a is acted on by a constant force F through a distance s, the force being in the same direction as the displacement s. In the process

---

* *Equation 20 is however applicable to alternating current devices which are purely 'resistive', like electric fires.*

an energy *Fs* (see equation 15) is transferred to the puck; we say the puck acquires a kinetic energy (written K.E.), i.e. energy of motion, of *Fs*. This energy is also the area under the force-distance curve, as shown in Figure 14b. We can express the energy transfer differently by recalling our primary definition of a force (equation 14 of Unit 3), namely

$$F = ma \dots\dots\dots\dots(21)$$

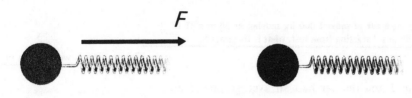

(*a*)

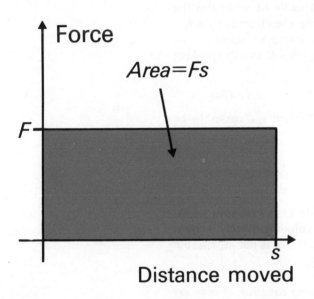

Distance moved

(*b*)

Figure 14 *Kinetic energy.* (*a*) *A puck of mass* m *is acted on by a force* F *through a distance* s. (*b*) *The energy transfer is shown as the shaded area, of magnitude* Fs.

So the K.E. acquired by the puck is given by

$$\text{K.E.} = \text{Area under force-distance}$$
$$\text{curve up to distance } s$$

$$= Fs$$

$$= ma\,s$$

i.e.
$$\text{K.E.} = m(as) \dots\dots\dots\dots(22)$$

But, under conditions of constant acceleration, there is a simple relation between the final velocity *v*, the initial velocity *u*, the acceleration *a*, and the distance gone, *s*.

Do you remember the relation?

If not, you should refer to Appendix 1, Unit 3; in particular see equation 7:

$$v^2 = u^2 + 2\,as \dots\dots\dots\dots(23)$$

Since the puck started from rest, i.e. $u = 0$,

$$v^2 = 2\,as \dots\dots\dots\dots(24)$$

or
$$as = \tfrac{1}{2}v^2$$

Substituting equation 24 into equation 22 gives

$$\text{K.E.} = \tfrac{1}{2}mv^2 \dots\dots\dots\dots(25)$$

33

Why, you may ask, did we go to all this trouble when we could simply have multiplied the force by the distance? In some situations this may indeed be possible, but if a moving object of mass m emerges from hiding with a velocity v, we have no idea what force acted on it nor through what distance it acted; yet we can determine its kinetic energy by applying equation 25.

SAQ 12
What is the kinetic energy of a car of mass 1 000 kg moving at 50 m s^{-1}? If the car took 20 s to reach 50 m s^{-1} starting from rest, what is its power?

The problem is worked out on p. 48.

The kinetic energy is 1.25×10^6 J and the car had an average power of 6.25×10^4 W, or 62.5 kW.*

It should perhaps be pointed out that all our discussions of energy transfers in mechanical systems have been non-relativistic, i.e. the objects have all been moving at speeds which are very much less than that of light. Thus, in equation 25, the mass m should really be written as the rest mass, m_0 (the mass when the velocity of the object tends to zero), while the measured velocity, v, is actually the improper velocity, v_{im}. Therefore the low-speed relation for kinetic energy should strictly speaking be written as

$$\text{K.E.} = \tfrac{1}{2} m_0 v_{im}^2 \quad \ldots \ldots \ldots \ldots (26)$$

But what, if anything, happens to equation 26 when v_{im} ceases to be negligible compared with the velocity of light?

4.4.5 Relativistic energy

While discussing the conservation of momentum in Unit 3 (section 3.6.3), it was pointed out that, as the velocities of the colliding objects increase towards that of light, what is conserved in the collision is not $m_0 v_{im}$, but

$$p = m_0 v_{im} / \sqrt{1 - v_{im}^2/c^2} \quad \ldots \ldots \ldots \ldots (27)$$

It was also pointed out that one way of interpreting equation 27 is to say that instead of remaining constant at m_0 the mass m of an object changes with speed v_{im} as follows:

$$m = m_0 / \sqrt{1 - v_{im}^2/c^2}$$

The mass m, you may remember, is called the relativistic mass of the object (section 3.6.3 of Unit 3). We might therefore be tempted to the conclusion that the relativistic expression for kinetic energy should be, not $\tfrac{1}{2} m_0 v_{im}^2$, but

$$\tfrac{1}{2} \left(\frac{m_0}{\sqrt{1 - v_{im}^2/c^2}} \right) v_{im}^2.$$

Such a conclusion would, however, be incorrect. To arrive at the correct expression for the kinetic energy of an object we must go back to first principles.

Can you remember how we derived the expression $\tfrac{1}{2} mv^2$, i.e. $\tfrac{1}{2} m_0 v_{im}^2$?

We multiplied the force F which acted on the object by the distance s through which it acted; this being the fundamental definition for the energy transferred to the object (equation 15), i.e. we evaluated the area

---

* *A unit of power that is sometimes used is the* horse-power. *1 horse-power = 746 W = 0.746 kW. It is evidently not an approved SI unit! The car in the example of* SAQ 12 *had a power of $6.25 \times 10^4/746 = 84$ horse-power.*

horse-power

under the graph of F plotted against s. We then assumed, and this is the point to note, that

$$F = ma \quad \ldots \ldots \ldots \ldots \ldots (21)$$

To spell it out, and it is worth spelling out, we assumed that if the body of mass m has an acceleration a at a particular instant, then the force acting on the body at that instant is the product of m and a. While this may appear to be a simple enough definition of the magnitude of the force F, equation 21 as it stands is a useless definition of force when the speed of the object approaches that of light; useless because it does not state whether the mass m is the value assigned by a stationary observer or by one travelling with the object. Equation 21 is also useless because it does not state the positions of the clocks used in measuring the speeds of the object and hence the rate at which that speed is changing. In other words, a worthwhile definition of force must state quite explicitly where all the observers (actually the clocks) are to be stationed.

It is perhaps interesting that the more general definition of force, the definition that may be applied unambiguously even for objects moving at speeds approaching that of light, can be traced back to Newton. To see how it is arrived at, we first rewrite equation 21 in a somewhat different form as follows. This re-arrangement will be carried out in the low-speed limit where the questions about where the clocks are located do not arise.

From equation 21 $\qquad F = ma \quad \ldots \ldots \ldots \ldots (21)$

But from the definition of acceleration as the rate of change of velocity

$$a = \delta v / \delta t$$

where δv is the change in velocity that takes place in a time δt.

I.e. $\qquad F = m\, \delta v / \delta t \quad \ldots \ldots \ldots \ldots (28)$

Assuming tentatively that m is constant, $m\delta v$ just means the change in the value of mv. Since mv is the momentum p of the object, $m\delta v$ just measures δp, the change in the value of p*.

Substituting δp for $m\delta v$ in equation 28 gives

$$F = \delta p / \delta t \quad \ldots \ldots \ldots \ldots (29)$$

Equation 29 was actually Newton's own definition of force, namely the change δp in the momentum of a body divided by the time interval δt during which this momentum change occurred. Since Newton's clocks were fixed in the laboratory the time interval δt in equation 29 is, in our terminology, an improper time interval δt_{im}. Therefore his definition of a force was

$$F = \delta p / \delta t_{im} \quad \ldots \ldots \ldots \ldots (30)$$

To Newton, p was simply mv (or $m_0 v_{im}$ in our notation). To us the correct definition of p is always

$$p = m_0 v_{im} / \sqrt{1 - v^2_{im}/c^2} \ldots \ldots (31)$$

Taken together, equations 30 and 31 provide a working definition of the force acting on a body. You will not be expected to remember these equations but, given equations 30 and 31, you may be able to describe how they could be employed to calculate the force acting on a body.

---

* As an example suppose the velocity of a 2 kg puck changes from 11.0 m s⁻¹ to 11.7 m s⁻¹. The change in momentum δp is the final momentum minus the initial momentum, i.e. (2×11.7) kg m s⁻¹ $-(2 \times 11.0)$ kg m s⁻¹ $= 23.4$ kg m s⁻¹ $- 22.0$ kg m s⁻¹ $= 1.4$ kg m s⁻¹.) The corresponding value of $m\,\delta v$ is $2(11.7 - 11.0)$ kg m s⁻¹ $= 1.4$ kg m s⁻¹, which is identical to δp.

> You are in a 'laboratory' situated somewhere in outer space. A space ship
> 'burns' past your laboratory. How would you set about deducing the force
> being provided by the ship's engines? Assume that the passengers on the space
> ship have told you their assessment of the mass m_0. Assume also that m_0
> remains constant throughout (i.e. that the total mass of the ship is very much
> greater than that of the fuel burnt in the engine).

You must measure the momentum (as defined by equation 31) at one instant and then measure it again a time δt_{im} later. Dividing the change in the momentum δp by the time interval δt_{im} gives the required force (equation 30). To determine the momentum you must, as equation 31 shows, determine the improper speed of the ship; this involves placing a couple of clocks a known distance apart in the 'laboratory'. Dividing the separation between the clocks by the difference in their time readings taken as the ship passes each clock gives the improper velocity.

> How would you, situated in your laboratory, calculate the energy which has
> been transferred from chemical form in the fuel into kinetic energy, while the
> rocket moves through a distance s as measured in your laboratory? (Assume
> s to be in the same direction as that in which the rocket moves.)

As we have seen (equation 15), the energy transferred is defined as being 'force × distance' or equivalently as being the area under the appropriate section of the force-distance curve. To calculate the energy transfer, you would have to calculate F as before, using equations 30 and 31. Then you would plot a graph showing how this calculated value of F varied as the ship moved through the distance s measured in the laboratory. Finally, you would evaluate, perhaps by 'counting squares' the area under the curve covering the distance s. While such a calculation can always be made from first principles each time it is required, it is convenient to have a once-for-all expression.

You saw, in section 4.4.4, that it is easy to work out the kinetic energy, $\frac{1}{2}m_0 v_{im}^2$, when the body is moving slowly enough for the simple formula $F=ma$ to apply.

The calculation is more difficult, however, when force is defined by equations 30 and 31. If you are interested, and know a little calculus, you will find the calculation worked through in Appendix 2 (Black). We shall just quote the result here:

$$\text{K.E.} = \gamma m_0 c^2 - m_0 c^2 \dots\dots\dots(32)$$

where
$$\gamma = 1/\sqrt{1 - v^2_{im}/c^2} \dots\dots\dots(33)$$

The symbol γ (pronounced 'gamma') is introduced simply to save the trouble of writing down $1/\sqrt{1 - v^2_{im}/c^2}$ which, as you may have noticed, is a quantity and keeps cropping up whenever we have to do with relativistic motion, that is with motion in which the improper velocity, v_{im}, is not negligible compared with the velocity of light, c.

Note, in passing, that equation 31 in Unit 3 (section 3.6.3) can be written in a conveniently short form, using γ:

$$m = m_0/\sqrt{1 - v^2_{im}/c^2} \dots\dots(31)$$
$$\text{(Unit 3)}$$
$$= \gamma m_0 \dots\dots\dots\dots(34)$$

So the kinetic energy could also be written as

$$\text{K.E.} = mc^2 - m_0 c^2 \dots\dots\dots(35)$$

in which we have replaced γm_0 by m.

Before we discuss further the implications of the result expressed in equations 32 or 35, it would be as well to check whether this formula for the kinetic energy can be reduced to the more familiar formula $\frac{1}{2}m_0 v^2{}_{im}$, when $v_{im} \langle\langle c$.

This amounts to finding an approximate value for

$$\gamma = 1/\sqrt{1 - v^2{}_{im}/c^2} = (1 - v^2{}_{im}/c^2)^{-\frac{1}{2}}$$

when $\qquad\qquad v_{im} \langle\langle c.$

You have come across this sort of problem before—refer back to Unit 3, section 3.3.1, equation 3, to refresh your memory. Here it is again:

$$(1 + x)^m \approx 1 + mx + \quad \ldots\ldots\ldots\ldots (3)$$
$$\text{(Unit 3)}$$

provided that $x \langle\langle 1$.

In the present case we have $x = -v^2{}_{im}/c^2$, which is indeed very much smaller than unity if $v_{im} \langle\langle c$, and we have $m = -\frac{1}{2}$.

So $\qquad (1 - v^2{}_{im}/c^2)^{-\frac{1}{2}} = 1 + \frac{1}{2}(v^2{}_{im}/c^2) + \ldots$

Put this approximate value for γ into equation 32:

$$\text{K.E.} \approx (1 + \frac{1}{2}v^2{}_{im}/c^2)m_0 c^2 - m_0 c^2$$
$$\approx \frac{1}{2}v^2{}_{im}/c^2 \cdot m_0 c^2$$
$$\approx \frac{1}{2}m_0 v^2{}_{im}$$

So the usual, non-relativistic formula is an approximate form of the more general, relativistic formula (equation 32). The approximation is a very good one at low velocities.

Now look again at equation 35.

Equation 35 tells us that kinetic energy—energy associated with motion relative to the observer—is made up of some amount of energy, (mc^2), minus some other amount of energy $(m_0 c^2)$. Let's turn equation 35 around a bit.

$$mc^2 = m_0 c^2 + \text{K.E.} \quad \ldots\ldots\ldots\ldots (36)$$

This tells us that some amount of energy (mc^2) is made up of some other amount of energy $(m_0 c^2)$, which does *not* depend on the motion of the object, plus the kinetic energy which *does* depend on the motion of the object—it is zero when the velocity v_{im} is zero.

This suggests strongly that the quantity mc^2 must be some sort of 'total energy', and that its two components are a 'rest-mass energy', $m_0 c^2$, and a kinetic energy:

rest-mass energy

'total (relativistic) energy' = 'rest-mass energy' + (relativistic) kinetic energy

$$mc^2 = m_0 c^2 + \text{K.E.} \quad \ldots\ldots\ldots\ldots (36)$$

Now look again at equation 34:

$$m = \gamma m_0$$

This tells us that when $v_{im} = 0$, which makes $\gamma = 1$, then $m = m_0$; to measure m, the 'rest-mass' we have to measure the mass of the object when it is at rest, or at least when it is moving so slowly that we can neglect v_{im} compared with c.

There is no difficulty about measuring the mass of a stationary, or quasi-stationary object, at least in principle. So we can, in principle, say what the 'rest-mass energy', $m_0 c^2$ is. If, then, we measure its improper velocity, v_{im}, we can work out its kinetic energy, using equations 32 and 33. And so we can work out the 'total energy', mc^2.

All this seems nice and neat and plausible, but does the quantity 'rest-mass energy' have any physical meaning?

It was Einstein who first guessed that if, in some process, matter of rest-mass m_0 were to 'disappear', the equivalent energy m_0c^2 should appear in its place.

The name of Hiroshima is a reminder that Einstein's guess was correct.

It is a small step, but conceptually a very important one, from saying

'in certain processes mass can get converted into an equivalent amount of energy',

to saying

'mass and energy are equivalent—they are two equivalent forms or modifications of a single attribute of matter'.

The latter is the point of view of contemporary physics. It is all the more firmly held because the disappearance or appearance of matter (energy) and the corresponding appearance or disappearance of the corresponding energy (mass) turns out to be an extremely commonplace occurrence.

You have already encountered one example of such a process in the TV programme of Unit 2, where you saw that muons disintegrate spontaneously and shoot out positrons in the process. The positrons get their energy (both kinetic energy and rest-mass energy) from the conversion of the rest-mass of the muon.

Indeed in *all* processes, physical and chemical, the appearance of energy is always accompanied by the corresponding disappearance of the equivalent amount of mass. But, unless the amount of energy that appears or disappears is quite large, the equivalent amount of mass is too small to be measurable; after all, c^2 is a very large number: 9×10^{16} m^2 s^{-2}.

To get an idea of the quantitative implications of m_0c^2, try the following problems:

The rest-mass of the debris after an atomic bomb explosion has been estimated to be some 10^{-3} kg less than the initial rest-mass of the materials. How much energy is released in such an explosion?

The energy released is the rest-mass 'destroyed' $\times c^2$, or $10^{-3} \times 9 \times 10^{16}$ kg m^2 s^{-2}

$= 9 \times 10^{13}$ J (since 1 J $=$ 1 kg m^2 s^{-2}).

Most of the energy will be released as heat, which, as you will learn in Unit 5, is none other than the kinetic energy of the atoms and molecules of all the matter in the neighbourhood of the explosion.

Probably the amount of energy 9×10^{13} J doesn't mean anything to you. An order of magnitude estimate in terms of roast chickens might help.

If a 15 kW domestic cooker takes 2 hours to roast a chicken, make an order of magnitude estimate of how many chickens might be roasted in the heat of such an atomic explosion.

Assuming the cooker is on continuously, it will be using 15 000 J of energy every second (15 kW $=$ 15 000 W and a watt is a joule per second).

So in 2 hours, the cooker uses $15 \times 10^3 \times 2 \times 60 \times 60$ J, or about 10^8 J. Since the energy released in the explosion is about 10^{14} J, it follows that the number of chickens that could be roasted with that amount of energy is $10^{14}/10^8 = 10^6$—a 'mega chicken'.

It follows, of course, that the relatively modest amounts of energy given out in ordinary chemical reactions, of the kind you will be doing in your home experiments—a few joules at most—involve negligible mass changes. This is because chemical reactions depend on electromagnetic forces which are very much weaker than nuclear forces, and so the energies liberated in chemical reactions are characteristically some millions of times less than those liberated in nuclear reactions.

This vast amount of energy derived from a mass loss of only 1 gm.

4.5 Recapitulation

Surveying the basic forces, in terms of which all macroscopic forces can be explained, we looked in turn at gravitational, electromagnetic and nuclear interactions. We saw how the force F of gravitational attraction between spherical masses depended on the value of the masses m_1 and m_2 and on their separation r apart. In particular we discovered that (equation 4)

$$F = G m_1 m_2 / r^2$$

where G is determined experimentally.

Having found that when electric currents flow in adjacent metallic wires the two wires interact, we used this phenomenon as a means of defining the magnitude of the current. We then described an experiment in which known amounts of electric charge could be collected on metal spheres. A study of how the electrostatic force F between two spheres possessing charges Q_1 and Q_2 depended on their separation r apart showed that (equation 5)

$$F = \left(\frac{1}{4\pi\epsilon_0} \right) Q_1\, Q_2 / r^2$$

where $1/4\pi\epsilon_0$ is determined experimentally.

Although we discussed electromagnetic forces in terms of magnetic and electrostatic forces, we pointed out that both are due to electric charges, the magnetic force arises when the charges move relative to the observer; electrostatic forces arise when the charges are at rest relative to the observer. It was illustrated how the conceptual difficulty of 'action at a distance' could be avoided if fields were introduced. As an important example of a field, the electric field at a point distant r from a charge Q is (equation 7)

$$E = \left(\frac{1}{4\pi\epsilon_0} \right) Q / r^2$$

On looking into what happened whenever a force moved its point of application, we were led to think about the role of fuels and so to the concept of energy. We saw that a sensible measure of the energy transferred in a mechanical process is given by the area under the graph showing how the applied force varies with distance. This result was then applied by calculating potential energies, a particularly important example of which is the potential energy P.E. of a charge Q_1 at a distance R from another charge Q_2. It was shown (equation 17) that

$$\text{P.E.} = \left(\frac{1}{4\pi\epsilon_0} \right) Q_1\, Q_2 / r$$

The kinetic energy of a body was shown to be $\frac{1}{2} m_0 v_{im}^2$ so long as $v_{im} \ll c$. Here the rest-mass, m_0, is mass determined at low speeds. Examining the general expression for the kinetic energy of an object, we found that the mass of an object increases as its kinetic energy increased. This expression also suggested that an amount of energy $m_0 c^2$ would appear if a mass m_0 of matter were to disappear.

This led us to the idea of equivalence of mass and energy.

Appendix 1

Electrical potential energy

The area under the curve shown in Figure 12 (p. 29) between R and infinity represents the electrical potential energy (P.E.) of Q_1 when it is a distance R from Q_2; the energy transferred in a process being the area under the appropriate section of the force-distance curve.

i.e.

$$\text{P.E.} = \text{Area under the force-distance}$$
$$\text{curve between infinity and } R$$

$$= \int_R^\infty F \, dr$$

But from equation 5, $F = Q_1 Q_2 / 4\pi\epsilon_0 r^2$.

Therefore:

$$\text{P.E.} = \int_R^\infty \frac{Q_1 Q_2}{4\pi\epsilon_0 r^2} \, dr$$

Taking the constant terms Q_1, Q_2, and $1/4\pi\epsilon_0$ outside the integral gives

$$\text{P.E.} = \frac{Q_1 Q_2}{4\pi\epsilon_0} \int_R^\infty \frac{dr}{r^2}$$

$$= \frac{Q_1 Q_2}{4\pi\epsilon_0} \left[-\frac{1}{r} \right]_R^\infty$$

∴

$$\text{P.E.} = \frac{Q_1 Q_2}{4\pi\epsilon_0} \left[0 - \left(-\frac{1}{R} \right) \right]$$

i.e.

$$\text{P.E.} = \frac{Q_1 Q_2}{4\pi\epsilon_0 R}$$

Appendix 2 (Black)

Relativistic kinetic energy

As a result of the discussion in section 4.4.5, we agreed to define the force F acting on a body of rest-mass m_0, moving at a speed v_{im}, as

$$F = \frac{dp}{dt_{im}} \quad \dots\dots\dots\dots\dots\dots\dots(37)$$

where

$$p = \frac{m_0 v_{im}}{\sqrt{1 - (v^2_{im}/c^2)}} \quad \dots\dots\dots\dots(31)$$

Equation 37 is equation 30 taken to the limit as δt_{im} tends to zero.

To calculate the energy transferred to a body of rest-mass m_0 as it is accelerated from rest to a final velocity v_{tm}, corresponding to a final momentum p, we must calculate the appropriate area under a graph showing how the force acting on the body changes with distance, x, as measured by a stationary observer. Using the notation of calculus, the resulting energy of the body, its kinetic energy after it has moved through a distance, x, may be written as

$$\text{K.E.} = \int_0^x F \, dx$$

where F is defined by equations 37 and 31.

i.e.
$$\text{K.E.} = \int_0^x \frac{dp}{dt_{im}} \, dx$$

Rewriting the term inside the integral as $dp \, (dx/dt_{im})$ gives

$$\text{K.E.} = \int_0^p dp \, \frac{dx}{dt_{im}}$$

where p is the final momentum of the object.

But $dx/dt_{im} = v_{im}$

$$\therefore \quad \text{K.E.} = \int_0^p v_{im} \, dp$$

Integrating by parts, we obtain

$$\text{K.E.} = \left[v_{im} \, p \right]_0^{v_{im}} - \int_0^{v_{im}} p \, dv_{im}$$

Substituting for p from equation 31, and also expressing $v \, dv$ as $\tfrac{1}{2}d\,(v^2)$, gives

$$\text{K.E.} = \left[\frac{m_0 \, v_{\text{im}}^2}{\sqrt{1-(v^2_{\text{im}}/c^2)}} \right]_0^{v_{\text{im}}} - \frac{m_0}{2} \int_0^{v_{\text{im}}} \frac{d(v^2)}{\sqrt{1-(v^2_{\text{im}}/c^2)}}$$

The second term is a standard form. Integrating it, we have

$$\text{K.E.} = \left[\frac{m_0 \, v_{\text{im}}^2}{\sqrt{1-(v^2_{\text{im}}/c^2)}} \right]_0^{v_{\text{im}}} + \left[m_0 \, c^2 \sqrt{1-(v^2_{\text{im}}/c^2)} \right]_0^{v_{\text{im}}}$$

$$= m_0 \, c^2 \left[\frac{v_{\text{im}}^2/c^2}{\sqrt{1-(v^2_{\text{im}}/c^2)}} + \sqrt{1-(v^2_{\text{im}}/c^2)} \right]_0^{v_{\text{im}}}$$

$$= m_0 \, c^2 \left[\frac{1}{\sqrt{1-(v^2_{\text{im}}/c^2)}} \right]_0^{v_{\text{im}}}$$

$$\therefore \quad \text{K.E.} = \frac{m_0 \, c^2}{\sqrt{1-(v^2_{\text{im}}/c^2)}} - m_0 \, c^2$$

the general expression for the kinetic energy of a body of rest-mass m_0 moving at a velocity v_{im}.

Self-Assessment Questions

Section 4.2.1

Question 1 (*Objective 2*)

Make an order-of-magnitude estimate of the force of gravitational attraction between two adult human beings when standing side by side. Using a relation derived for spheres is bound to make the answer suspect; guesses of human masses correct to a factor of two or so will therefore be quite adequate.

Section 4.2.2

Question 2 (*Objective 3*)

Tick those statements below which correctly describe the nature of the electrodynamic interaction between two current-carrying wires.

1   The wires only interact if there is a current in both wires.
2   The wires will interact with a current in one wire alone.
3   The force between the wires is attractive when the currents are in opposite senses in each wire.
4   The force between the wires is attractive when the currents are in the same sense in each wire.
5   The forces decrease as the number of batteries in the circuits is increased.
6   The force acting on the wires is independent of the medium in which they are located.
7   The forces increase as the wires are brought closer together.

Question 3 (*Objective 4*)

Make an order-of-magnitude calculation of what would be the force of electrostatic attraction between two adults standing side by side if 1.5 C of charge had been transferred from one person to the other. Compare this electrostatic force with the force of gravitational attraction between the persons.

Question 4 (*Objective 5*)

The normal British one-bar electric fire takes a current of about 4 A. How many electrons pass through such a fire each hour?

Question 5 (*Objective 5*)

How many electrons were transferred from one person to the other in problem *SAQ* 3? Express this gain or loss as a percentage of the total number of electrons in each person. (You will need to know that atoms have an average diameter of about 10^{-10} m and that in living matter each atom contains an average of about 10 electrons.)

Section 4.3.2

Question 6 (*Objective 6*)

Calculate the electric field at a distance of 2×10^{-10} m from a charge of 1.6×10^{-19} C.

Question 7 (*Objective 6*)

What will be the force on a 1.6×10^{-19} C charge when introduced into a region where the electric field is 3.6×10^{10} N C^{-1}?

Section 4.4.2

Question 8 (*Objective 7*)

Make an order-of-magnitude calculation of the (chemical) energy required to push a car by hand through a distance of 1 mile. Assume the car is in 'neutral'. (As a clue, first try and decide, to the nearest power of ten, what force is required to keep a car moving.) Ignore the extra effort required to get the car going.

Section 4.4.3

Question 9 (*Objective 8*)

What energy is required to bring a charge of -1.6×10^{-19} C up from infinity to a distance of 2.5×10^{-10} m from a charge of 1.6×10^{-19} C?

Question 10 (*Objective 8*)

The potential difference between the 'live' and the neutral terminals on a mains plug is typically 250 V in Britain. When connected to such mains, a one-bar electric fire takes a current of 4 A. How much thermal energy is transferred to the room per second by such a fire?

Question 11 (*Objective 9*)

What current passes through a 150 W light bulb operating from a 250 V direct current supply?

Section 4.4.4

Question 12 (*Objective 8*)

What is the kinetic energy of a car of mass 1 000 kg moving at 50 m s^{-1}?
If the car took 20 s to reach 50 m s^{-1} starting from rest, what is its power?

Self-Assessment Answers and Comments

Question 1

Assume the two adults have identical masses of 80 kg (i.e. about 13 stone) and that, when close together, their distance apart, between 'centres' will be about 0.3 m (i.e. about 1 ft.). Substituting $m_1 = m_2 = 80$ kg and $r = 0.3$ m into equation 4 (a relation, remember, which was derived using spheres) gives the force of attraction, F, as

$$F = G\, m_1 m_2 / r^2$$

$$= \frac{6.67}{10^{11}} \times \frac{80^2}{0.3^2} \frac{\text{N m}^2}{\text{kg}^2} \frac{\text{kg}^2}{\text{m}^2}$$

substituting $G = 6.67 \times 10^{-11}$ N m^2 kg^{-2}.

Because of the uncertainties inherent in applying equation 4 to non-spherical masses there would be little point, for example, in using 'log tables' to evaluate the answer—rough approximations will be quite adequate.

$$F \approx \frac{7}{10^{11}} \times \frac{6.5 \times 10^3 \text{ N}}{10^{-1}}$$

$$\approx \frac{4.5}{10^6}$$

$$F \approx 10^{-5} \text{ N}$$

The reason for deciding to approximate 4.5×10^{-6} N to 10^{-5} N, rather than to 10^{-6} N, is that the calculation has been made by multiplying and dividing quantities. To 'convert' 5×10^{-6} to 10^{-5} requires only a factor of 2 'wrong', but to convert it to 10^{-6} requires a factor of 5.

Question 2

The correct statements are 1, 4, 7. See section 4.2.2.

Question 3

Substituting $Q_1 = Q_2 = 1.5$ C and $r = 0.3$ m (say) into equation 5 gives the attractive force F as

$$F = \frac{Q_1 Q_2}{4\pi\epsilon_0 r^2}$$

$$\approx \frac{9 \times 10^9 \times (1.5)^2}{(0.3)^2} \frac{\text{N m}^2}{\text{C}^2} \frac{\text{C}^2}{\text{m}^2}$$

$$\text{taking } \frac{1}{4\pi\epsilon_0} \approx 9 \times 10^9 \text{ N m}^2 \text{ C}^{-2}$$

As equation 5 is really only applicable to spheres, rough and ready approximations can be made.

i.e.
$$F \approx \frac{10^{10} \times 2}{10^{-1}} \text{ N}$$

$$\approx 2 \times 10^{11} \text{ N}$$

$$\approx 10^{11} \text{ N}$$

Question 4

By 4 A is meant 4 C of charge per second passing a point in the wire. But there are $1/1.6 \times 10^{-19} = 6.25 \times 10^{18}$ electrons per coulomb of charge (the electronic charge is 1.6×10^{-19} C).

i.e.
$$4 \text{ A} = 4 \text{ C s}^{-1}$$

$$= 4 \text{ C} \times 6.25 \times 10^{18} \text{ electrons s}^{-1}$$

$$4 \text{ A} = 2.5 \times 10^{19} \qquad \text{electrons passing a point in the wire per second.}$$

∴ In 1 hour the number of electrons passing through the fire will be $2.5 \times 10^{19} \times 60 \times 60 \approx 10^{23}$.

Question 5

Since the electronic charge is 1.6×10^{-19} C there are $1/1.6 \times 10^{-19} = 6.25 \times 10^{18}$ electrons per coulomb. Therefore the number of electrons in the 1.5 C charge, transferred from one person to another, is $1.5 \times 6.25 \times 10^{18} \approx 10^{19}$.

To estimate the total number of electrons in a human body we estimate the number of atoms in the body and then multiply the result by 10; there is an average of about 10 electrons per atoms in living matter.

Treating an atom as a 10^{-10} m cube it will have a volume of $(10^{-10})^3$ m^3, i.e. 10^{-30} m^3. A human's volume is about 10^{-1} m^3 (e.g. 1.5 m high $\times$ 0.2 m wide $\times$ 0.3 m broad). Therefore the number of atoms in the body is about $10^{-1}/10^{-30} = 10^{29}$. Hence the number of electrons is about $10 \times 10^{29} = 10^{30}$. The gain or loss of the 10^{19} electrons expressed as a percentage of the total number of electrons in the body is

$$\frac{10^{19}}{10^{30}} \times 100 \text{ per cent}$$

$$= 10^{-9} \text{ per cent}$$

Put differently only 1 atom in 10^{11} would have to lose or gain an electron to produce the total change in charge.

Question 6

The electric field E at a distance r from a charge q is, from equation 7,

$$E = q/4\pi\epsilon_0 r^2$$

Substituting $q = 1.6 \times 10^{-19}$ C, $r = 2 \times 10^{-10}$ m, and $1/4\pi\epsilon_0 = 9 \times 10^9$ N m^2 C^{-2} gives

$$E = \frac{9 \times 10^9 \times 1.6}{10^{19} \times (2 \times 10^{-10})^2} \frac{\text{N m}^2}{\text{C}^2} \frac{\text{C}}{\text{m}^2}$$

$$= \frac{9 \times 1.6 \times 10^{10}}{4} \text{ N C}^{-1}$$

$$= 3.6 \times 10^{10} \text{ N C}^{-1}$$

Question 7

Since the electric field at a point is, by definition, the force per coulomb of charge placed at the point, it follows that the force F on a charge of 1.6×10^{-19} C at a point where the electric field is 3.6×10^{10} N C^{-1} is

$$F = (3.6 \times 10^{10}) \times (1.6 \times 10^{-19}) \, \frac{N}{C} \, C$$

i.e.

$$F = 5.76 \times 10^{-9} \text{ N}$$

Question 8

The sort of force we provide via our leg muscles in pushing the car is probably about the same as the force it takes to lift up say, a couple of stones of potatoes, again using our leg muscles. Now the weight of an object of mass 2×14 lb. $= 28$ lb. $= 28/2.2$ kg is 13×9.8 N $\approx 10^2$ N (the acceleration due to gravity being 9.8 m s^{-2}). Since 1 mile $\approx 1\,500$ m the energy supplied by the pusher is 'force $\times$ distance' $\approx 10^2 \times 1.5 \times 10^3 \approx 10^5$ N m, or 10^5 J.

Question 9

The energy required is by definition, the electrical potential energy of the -1.6×10^{-19} C charge when 2.5×10^{-10} m from the $+1.6 \times 10^{-19}$ C charge. From equation 17 the potential energy of a charge Q_1 when a distance R from Q_2 is given by

$$\text{P.E.} = Q_1 \, Q_2 / 4\pi\epsilon_0 R.$$

Here

$$Q_1 = -1.6 \times 10^{-19} \text{ C}$$

$$Q_2 = +1.6 \times 10^{-19} \text{ C}$$

$$R = 2.5 \times 10^{-10} \text{ m}$$

$$1/4\pi\epsilon_0 = 9 \times 10^9 \text{ N m}^2 \text{ C}^{-2}$$

$$\therefore \text{ P.E.} = - \frac{9 \times 10^9 \times 1.6 \times 1.6}{10^{19} \times 10^{19} \times 2.5 \times 10^{-10}} \, \frac{\text{N m}^2}{\text{C}^2} \, \frac{\text{C}^2}{\text{m}}$$

$$\text{P.E.} = -9.2 \times 10^{-19} \text{ J}$$

The minus sign indicates that energy is transferred from the electric field to the pusher in this case; the positive charge pulls in the negative one.

Question 10

The energy transferred as 1 C moves through a potential difference of 250 V, i.e. 250 J C^{-1}, is 250 J. (This follows from the definition of potential difference as the energy transferred per coulomb.) But a current of 4 A means that 4 C of charge pass through the force per second. Therefore the total thermal energy appearing in the fire per second is 250 J C$^{-1} \times$ 4 C s$^{-1} = 10^3$ J s^{-1}.

Question 11

Equation 20 shows that

$$\text{Power} = V \times I$$

i.e.

$$I = \frac{\text{Power}}{V}$$

In this problem the power $= 150$ W $= 150$ J s^{-1}, and $V = 250$ V $= 250$ J C^{-1},

$$\therefore \quad I = \frac{150}{250} \frac{\text{J s}^{-1}}{\text{J C}^{-1}}$$

i.e. $\qquad\qquad I = 0.6$ C s$^{-1} = 0.6$ A.

Question 12

The kinetic energy of an object of mass m moving at a speed v is $\frac{1}{2}mv^2$.
Substituting $m = 10^3$ kg and $v = 50$ m s^{-1} gives

$$\text{K.E.} = \frac{1}{2} \times 10^3 \times (50)^2 \quad \text{kg} \frac{\text{m}^2}{\text{s}^{-2}}$$

$$= 1.25 \times 10^6 \text{ N m}$$

$$= 1.25 \times 10^6 \text{ J}$$

As the car took 20 s to obtain the K.E. its average power (the rate at which it gained K.E.) was $125 \times 10^4/20 = 6.25 \times 10^4$ J s$^{-1} = 6.25 \times 10^4$ W

Acknowledgements

Grateful acknowledgement is made to the following sources for illustrations for this Unit:
NATIONAL PHYSICAL LABORATORY, Crown Copyright, Fig. 4; H. W. SULLIVAN LTD., Fig. 5.

Notes

S.100—SCIENCE FOUNDATION COURSE UNITS

1   Science: Its Origins, Scales and Limitations
2   Observation and Measurement

3   Mass, Length and Time
4   Forces, Fields and Energy

5   The States of Matter

6   Atoms, Elements and Isotopes: Atomic Structure
7   The Electronic Structure of Atoms

8   The Periodic Table and Chemical Bonding
9   Ions in Solution

10   Covalent Compounds

11
12   } Chemical Reactions

13   Giant Molecules

14   The Chemistry and Structure of the Cell

15
16   } Cell Dynamics and the Control of Cellular Activity

17   The Genetic Code: Growth and Replication
18   Cells and Organisms

19   Evolution by Natural Selection
20   Species and Populations

21   Unity and Diversity

22   The Earth: Its Shape, Internal Structure and Composition

23   The Earth's Magnetic Field

24   Major Features of the Earth's Surface
25   Continental-Movement, Sea-floor Spreading and Plate Tectonics

26
27   } Earth History

28   The Wave Nature of Light

29   Quantum Theory
30   Quantum Physics and the Atom

31   The Nucleus of the Atom
32   Elementary Particles

33
34   } Science and Society